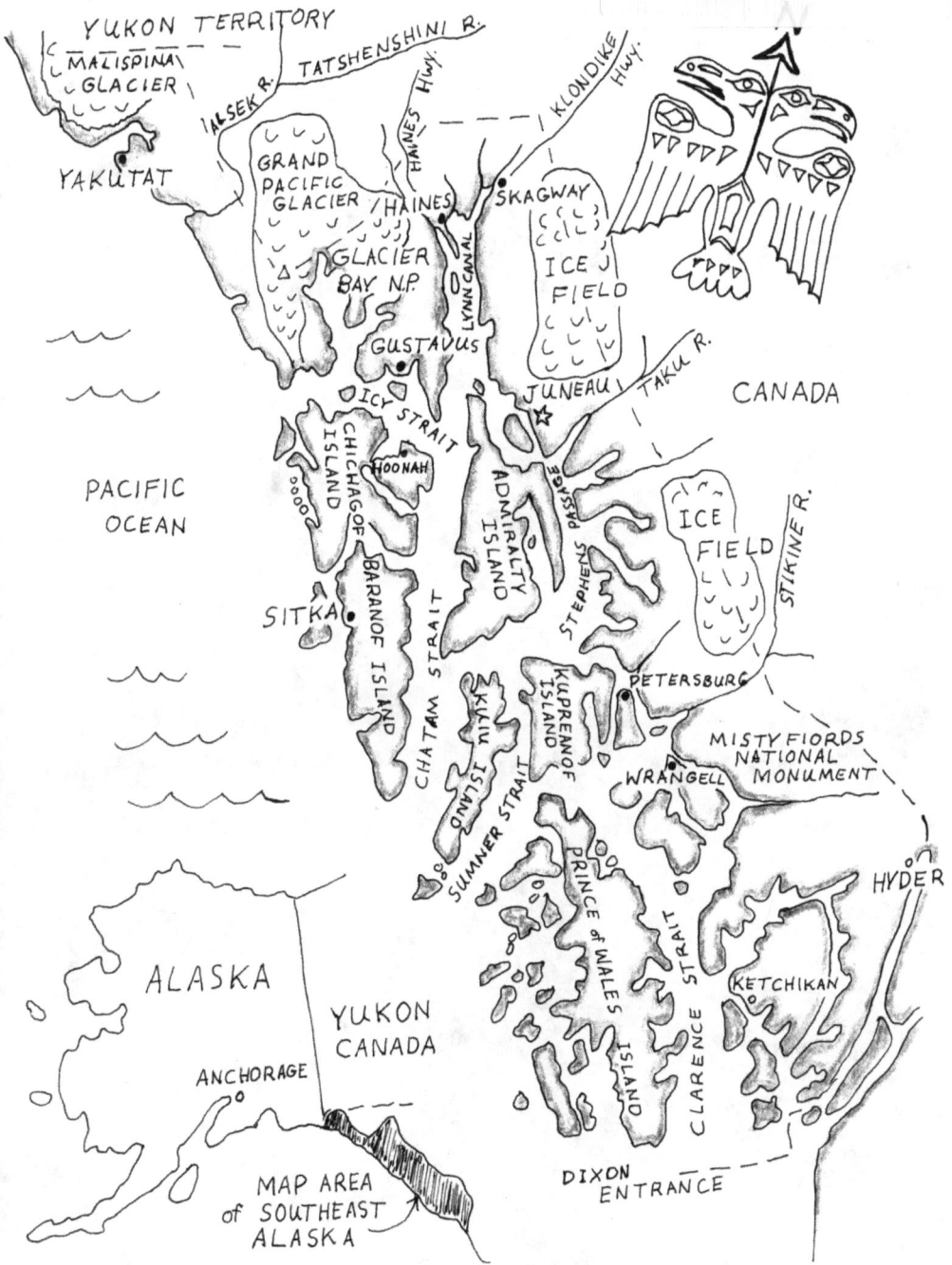

YUKON TERRITORY

MALISPINA GLACIER

TATSHENSHINI R.

ALSEK R.

HAINES HWY.

KLONDIKE HWY.

YAKUTAT

GRAND PACIFIC GLACIER

HAINES

SKAGWAY

GLACIER BAY N.P.

LYNN CANAL

ICE FIELD

GUSTAVUS

JUNEAU

TAKU R.

CANADA

ICY STRAIT

PACIFIC OCEAN

CHICHAGOF ISLAND

HOONAH

ADMIRALTY ISLAND

STEPHENS PASSAGE

ICE FIELD

STIKINE R.

SITKA

BARANOF ISLAND

CHATAM STRAIT

PETERSBURG

KUIU ISLAND

KUPREANOF ISLAND

WRANGELL

MISTY FIORDS NATIONAL MONUMENT

SUMNER STRAIT

HYDER

PRINCE of WALES ISLAND

CLARENCE STRAIT

KETCHIKAN

ALASKA

YUKON CANADA

ANCHORAGE

MAP AREA of SOUTHEAST ALASKA

DIXON ENTRANCE

Native Plants
of
Southeast Alaska

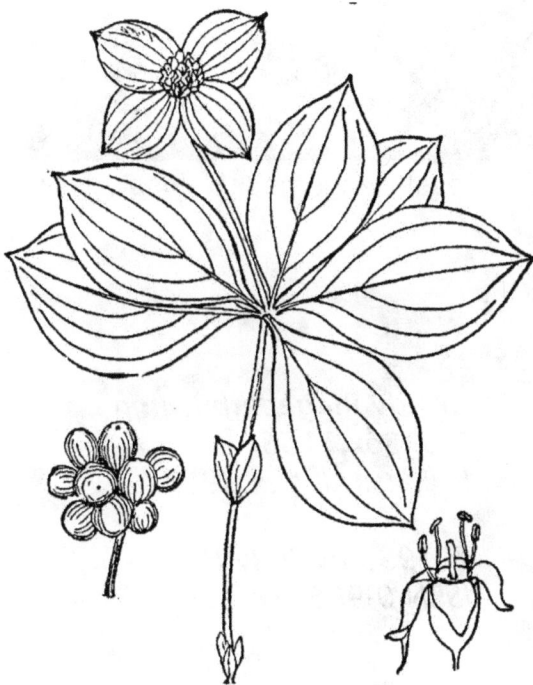

Judy Kathryn Hall

> In Wildness is the Preservation of the World
> Henry David Thoreau

Coast Range near Haines, Alaska

Windy Ridge Publishing
P.O. Box 1158
Haines, Alaska 99827

© 1995 by Judy Hall
ISBN 0-9658726-0-2

Cover photos by the author:Bog-Laurel, Black Cot-tonwood, Roseroot, Calypso Orchid and Indian-pipe.
Title page drawing: Bunchberry. Spline (aka spine):
Skunk Cabbage.

PREFACE

The idea for this field guide emerged during a Juneau Parks and Recreation hike to the John Muir Cabin near Juneau in 1986. Hikers expressed frustration identifying plants using guides written primarily for Interior Alaska. After discussions with friends, fellow naturalists and colleagues, I found enthusiastic agreement: a guide specializing in Southeast Alaska plants is needed. My years compiling information and keying plants began.

Experience for this endeavor came during my ten summers of ecological field work conducted in Oregon, Washington, and Southeast and South-central Alaska under the auspices of Oregon State University. Many voyages on the Alaska State Ferries took me to places all over our region to record flowering times, habitat and abundance information. That I lived in Juneau at the time--the center of the range of the book--was helpful. Extensive kayak trips allowed me to study beach, forest and bog plants.

The staff of the Alaska State Museum permitted me access to the herbarium developed by the Juneau Botanical Club, an organization of amateur botanists inspired by J.P. Anderson. The USFS Forestry Sciences Laboratory herbarium in Juneau proved invaluable for information on range, habitat and flowering times. Dr. Paul Alaback contributed to part of the introduction. I sincerely thank him for his sage advice and encouragement during this project. Collections made by John Thelinius and Wayne Robuck proved invaluable. The USFS Region 10 Ecology Program, provided important information on range extensions. Mary Muller of the USFS Chatham Area, John Delapp and Rob Lipkin of the Alaska Natural Heritage Program and the Nature Conservancy were especially helpful in providing information on other plant collections. Dr. Stanley L. Welsh graciously allowed my use of keys from ***Anderson's Flora of Alaska and Adjacent Parts of Canada***. Since this monumental book is out of print, it gave me further impetus to create this guide. Dr. Welsh is one of the botanical giants on whose shoulders I stood.

On numerous skiing and hiking trips, Bob Armstrong inspired me with his insight and encouragement. His experience as the author of **Birds of Alaska**, and coauthor of **The Nature of Southeast Alaska**, as well as his friendship, proved invaluable.

Claudia Kelsey not only played a large part in putting together the Juneau Botanical Club's herbarium but also put her creative hand to the task of drawing many of the illustrations. Her botanical and artistic talent, as well as our many hours, discussing botany at her home, I will always cherish. Terry Jacobson redrew illustrations for several trees and shrubs (and other plants), coming to my rescue when an illustration source changed from public domain status. He drew the Southeast Alaska map as well.

Julie Ross and her students at the University of Alaska SE field tested the manuscript and returned many helpful suggestions. Dan Berch also looked over the original manuscript and gave his able advice. Ellen Anderson meticulously edited the manuscript. Jane Ginter and Laura Lucas were my Pagemaker experts, and Steve Vick help me realize that someday I would understand the intricacies of InDesign. Also thanks to Tony Mecklenberg of Pt. Stevens Press and John Cloud of Time-Frame for their technical help.

Special thanks to Margaret Beilharz, a longtime friend, for a place to write at her home by a window where native Southeast Alaska plants outside inspired me. Special thanks to the late Lee Foster for friendship, encouragement and humor and for his technical support helping to power my computer by the central Alaskan sun. Thanks also to Danny Newby of Delta Junction for recovering and healing a crashed hard drive.

The revised edition would not have been possible without the advice and attention to detail supplied by Jim Green of Williwaw Publishing. A very special thanks goes to Terry Jacobson for his love, encouragement and faith in all our endeavors together.

Of course, thanks to Mom and Dad. Last--but not least--thanks to my son Sean Muir Alaback for putting up with mosquitoes inside his headnet during so many Alaskan hikes and for watching Sesame Street while I worked on this book.

Judy Hall Jacobson
Juneau, March 1995
Haines, October 2010 (revision)

Table of Contents

Table of Contents

Introduction

Southeast Alaska lies in the largest temperate rainforest on the planet. The forests, alpine ridges and complex wetlands of this wild land provide for a diverse and beautiful array of plant life. Profound changes driven by remote logging camps and other lands becoming towns, rapid increases in recreational use and natural resource development have come to this region. These changes accentuate the need for a simple, yet comprehensive account of the vascular plant species occurring in Southeast Alaska. It is my hope that this book will encourage residents, visitors, students, scientists, agencies, and developers alike to become more familiar with the flora of this beautiful region.

Species described in this manual grow in the Sitka Spruce--Western Hemlock biome. The range of this book extends from Dixon Entrance--the southern border of Alaska and Canada--north to Haines, Skagway, and Yakutat. Since the region extends west along Prince William Sound to Kodiak Island, this field guide is useful there as well. Approximately 1300 miles from end to end, this coastal land lies between latitudes 54 and 62 degrees north and 130 to 155 degrees west.

Species native to the area are the focus of this manual; however you will also find species that have been introduced and then naturalized covered as well. In an attempt to be comprehensive, some rare species or those growing barely within the range of the book are included.

How to use this guide
The Keys

Several approaches can be used to identify an unknown plant using this guide. One method is to follow the steps of the keys provided in each chapter. Keys enable the reader to reach more specific levels of plant identification. These are two-branched keys that rely mainly on vegetative rather than flower characteristics. Start with the ***Key to Families*** to identify the Group (fern or fern ally, conifer, or flowering plant--dicot or monocot) and Family of the plant in question. Then go to the beginning of the section on that Family to find the key to take you to genus and species. To conserve space, families and genera with less than five members do not have keys.

This method is most reliable for distinguishing between closely related species and is most useful for readers who are familiar with technical descriptions of plants. A glossary is included to explain technical terms used in the keys and text.

When a key describes flower characteristics and all that remains of the specimen is the fruit or flower of the previous year, use imagination and reconstruction to determine the way to go in the key. When confusion arises over which branch of the key best fits the specimen in hand, go both directions to see if one looks more plausible for your specimen than the other. Growing conditions, age, and season can cause plants of the same species to appear quite different.

Keep in mind that a key is our attempt to pigeonhole nature to make it understandable: a difficult task. Keys are static while nature is constantly changing. Use these keys as a guide to identification, not as the final word.

For readers less familiar with technical descriptions of plants, a faster, more enjoyable approach may be to use the "picture booking" technique. Find the illustration that most closely resembles the plant in question. Carefully compare the plant description with the actual plant. If the plant does not completely fit the description, check other species in the same genus or family.

Plant Descriptions

The arrangement of plants is in alphabetical order for family, genus, and species within each group. Two paragraphs cover common species and genera containing only one species. Information includes physical characteristics, geographic distribution, and comments on abundance and human use. The name of the genus and species of common plants stand alone in the text, above the description. Shorter, less detailed descriptions cover genera and species of less common plants. This format results in plants most likely encountered in the field to be more likely encountered in this guide. All species have illustrations.

Physical characteristics

Physical characteristics comprise the first paragraph of the plant description (or for less common species, the first part of the description). Description of the overall plant is first; then leaf, flower and fruit characteristics. This project began when it seemed inevitable the United States would adopt the metric system, which is yet to occur. This field guide uses the metric system.

Geographic distribution

The second paragraph of the description, (or for less common species, the last part of the description) begins with information on geographic distribution. Statewide range information is most accurate for Southeast Alaska. Plant ranges change as plants move into new areas, and botanical studies continue around the state.

Abundance

Information on abundance or likelihood of encountering a specific species helps in identifying plants. Species listed as "common" exist throughout its range. Species listed as "occasional" will be seen from time to time. Species listed as "uncommon" or "rare" are those that will be sure to raise a stir with their finder. Flowering times, though, can only be described in general terms because of plant sensitivity to year-to-year climatic variation, microclimate, habitat and elevation.

Human use

Human uses of plants are interesting to many people. If foraging, never take all plants in an area. Leave some for others to enjoy as well. Readers are encouraged to forage only after becoming thoroughly familiar with all poisonous plants in their area. Foragers should be particularly cautious with members of the Carrot and Buttercup families. Water Hemlock and Baneberry are deadly poisonous and very common, especially in coastal meadows and fields. The author has tested most species for edibility, but it is up to the user of this information not to consume any plant he or she has not unquestionably identified. Check references listed in the bibliography for more information on edible and medicinal plants.

Taxonomy

Where practical, a simple taxonomic treatment was adopted to aid species identification. Aside from ferns, I chose to use more recently developed taxonomic treatment than that used in previous Alaska floras (Hultén 1968, Welsh 1974). Rather than the six or more families employed by some taxonomists, the ferns are placed in three easily distinguishable families. This choice was made to simplify identification. The primary reference for taxonomy came from *A synonymized checklist of the vascular flora of the United States, Canada, and Greenland (Kartesz and Kartesz 1980).*

Nomenclature

Nomenclature follows that of *Kartesz and Kartesz (1980)* except where more recent and well-accepted revisions of taxa have occurred. At the end of the plant description, taxonomic synonyms used by *Hultén (1968)* or *Hitchcock and Cronquist (1973)* are referenced. One common name is used for each species to standardize common names. Names chosen are usually those used by *Muller (1982), Welsh (1974)* or *Hitchcock and Cronquist (1973)* in that order.

Taxonomists continue to revise plant names making an up-to-date field guide difficult to impossible. Every effort has been made to cross-reference other scientific names and those no longer in use. Former names and synonyms are italicized in the index.

The Illustrations

Line drawings rather than photographs have been chosen to illustrate the plants to attain a generalized depiction of each species and to highlight characteristics for use in identification. Several of the illustrations were drawn by Jeanne R. Janish and John H. Rumely and selected with permission by the University of Washington from the five volumes of *Vascular Plants of the Pacific Northwest (Hitchcock et al. 1955-1959).* Other drawings are from the British Columbia field guide series published by the British Columbia Provincial Museum. Stanford University granted permission for use of illustrations from *Flora of Alaska and Neighboring Territories (Hultén, 1968)*. Claudia Kelsey of Juneau drew several of the illustrations from herbarium and live specimens. Terry Jacobson of Haines drew others while many illustrations are from the **USDA-Natural Resources Conservation Service Plants Database** including **An Illustrated Flora of the Northern United States, Canada and the British Possessions. (Britton, N.L., and A. Brown. 1913), C. Kenneth Pearse, A. Parela** and **Wetland Flora: Field Office Illustrated Guide to Plant Species.**

Environmental Influences in Southeast Alaska
Geology

The southeast coast of Alaska is a natural geographic unit isolated from the interior by the Coast Range Mountains. This mountain range affects the distinctive local landforms, soils and climatic conditions resulting in the diverse alpine, bog and forest flora of the region. Many species are also common in the harsh climates of north and south coastal Alaska.

Plate tectonics greatly impacts the landscape of our region. At this moment, a chunk of earth termed the Yakutat terrain is sliding past and partially diving under northern Southeast Alaska. This process is pushing up the peaks of the St Elias Range of which the Chilkat Mountains are a part. The Pacific Plate is moving counter-clockwise west and north, and is jamming against the North American plate. Meanwhile, the North American plate is moving west. At the contact point lie fault lines: places where rocks shear. Geologists have determined that over geologic time, parts of Southeast Alaska moved along on the Pacific plate from places as far away as Tasmania.

Due to climate change, icefields and glaciers in the coastal mountains of Southeast Alaska have undergone rapid melting in the last 250 years. In response to the loss of staggering amounts of glacial ice, the earth's crust is rebounding here.

Researchers with the University of Alaska's Geophysical Institute have measured the rate of this rise at 30 millimeters (1 1/2") per year in Glacier Bay. This process, known as isostatic rebound, results in new emerging land and, therefore, impacts plants.

Considering the processes of plate tectonics and isostatic rebound are occurring simultaneously, the question emerges: is uplift due to isostatic rebound or is it caused by plate tectonics? The Queen Charlotte-Fairweather fault system, larger than the San Andreas fault, cuts through northern Southeast Alaska. Researchers say this strike-slip fault cuts too cleanly through northern Southeast Alaska for tectonic motion to be the main driving force behind uplift. Most likely it is melting ice that causes the land's rapid rise in elevation.

Climate

The climate of southern Alaska is relatively mild, despite its high latitude and the presence of the largest concentration of glacial ice of any nonpolar area in the northern hemisphere. Extensive ice fields in this region are primarily the result of heavy winter precipitation and cool summer temperatures rather than harsh winter climate.

Cold air pouring down from ice fields and large glaciers dramatically influence local plant distribution, sometimes allowing true alpine species to occur near sea level. From Dixon Entrance north to Prince William Sound, there is a steady decrease in growing season temperature and a decline in plant variety and robustness.

Low-pressure weather cells forming around 60 degrees latitude and warm ocean currents influences the weather in this region. The result is mild temperatures and heavy rain in lower elevations year round. In Southeast Alaska, rainfall averages 150-750 cm (60-300 inches) per year at sea level. Rainfall increases to 1000 cm (400 inches) per year at higher elevations, especially on exposed south- and east-facing slopes. At Skagway, rainfall drops to 640 mm (27 inches) per year.

In this land of heavy rain, wet takes on a different meaning. The "driest" habitats for plants occur on rocky windswept ridges and in sandy glacial outwash areas. Wet sites include areas continually bathed in water percolating down slope or where water has been impounded by glacial till or hard pan soils.

The south coastal region of Alaska is a crossroads between the flora of the Pacific Northwest (coastal British Columbia, Washington, and Oregon) and the boreal zone of Canada, interior Alaska, and Siberia. Approximately 75% of native plant species in coastal Alaska occur in the Pacific Northwest. Many species reach their northern limit of range in this region. It is also the southern limit for many boreal and Asiatic species. Biogeographers believe the Bering Land Bridge, which united Alaska and Siberia several times during the geologic past, was a critical factor in plant colonization of Alaska and partially explains the circumpolar distribution of many Alaskan species.

Major drainages such as the Taku, Stikine, and Alsek Rivers, and the head of Lynn Canal, cut through the massive batholiths of the Coast Range. These serve as migration corridors for plants from the Interior and are excellent places to see unusual species. For example, several alpine species and the uncommon subalpine fir occur at sea level in the Taku River area. Many additions to our flora not described in previous texts are from collections in these botanically unique areas.

Due to mild temperatures and abundant rainfall, forests are the dominant vegetative type in south coastal Alaska. Sitka Spruce and Western Hemlock over 700 years old and over six feet in diameter grow in large protected river valleys. More commonly though, trees are younger and smaller due to blow down caused by periodic wind storms or due to poorly drained infertile soils.

Dense forests of Western Hemlock and Sitka Spruce cover the land from the Oregon, Washington, and British Columbia coasts, extending 1,440 km (900 miles) along the Alaska coast from Dixon Entrance to Cook Inlet and Kodiak Island. Western Hemlock is the most common tree. On alluvial valley bottoms, along beaches, on edges of avalanche tracks, on steep slopes, and in the alpine, Sitka Spruce may dominate the forest. South of Port Frederick Sound, Western Red Cedar is widespread on all but the best drained and most fertile soils.

Mountain Hemlock is not necessarily a subalpine tree in our region. It is common on poorly drained soils in forests, near lakes, in peatlands, or on better-drained sites just below treeline. A variety of Lodgepole Pine, known locally as Shore Pine, also grows in peat bogs. Alaska Yellow Cedar, another tree of bogs, drops out at Prince William Sound, but is occasionally dominant in forests. For more than a century, though, large swaths of Yellow Cedar in the Tongass National Forest have been dying. Yellow Cedar staves off bugs and decay better than other trees in the Tongass, but a thin spring snowpack has left its roots susceptible to freezing. The decline of Yellow Cedar occurs at low elevations, which are experiencing less snow as climate warms. Red Alder is common along streams, beach fringes and on soils recently disturbed by logging. Black Cottonwood grows on the flood plains of major rivers and recently deglaciated areas on the mainland.

The greatest number of species and habitats occur in peat bogs known colloquially as muskegs. Peat (decayed plant matter, usually Sphagnum and sedges) compose bogs or poorly drained areas. Due to high soil acidity and high water tables, plant decomposition is slow, and nutrients are mostly unavailable to plants. Plants develop specialized adaptations to grow in such nutritionally impoverished conditions. Sundew and Bladderwort, common bog plants, derive nutrients, mainly nitrogen, by digesting insects trapped on their sticky leaves. Sitka Spruce survive in bogs by growing on logs buried beneath the moss. Several unique herbaceous species occur in pit ponds or seasonally flooded areas.

Many bog plants such as Arctic Starflower (Trientalis arctica) occur in the alpine as well. Most open boggy or sedge meadows called by locals alpine meadows are technically subalpine meadows created by heavy snow accumulation but which have relatively mild summer temperatures. Only on massive, rocky, windswept or ice-influenced ridges are true alpine plants such as Moss Campion (Silene acaulis) likely to grow.

Botanical mysteries abound in Southeast Alaska. Of interest are mountains known as nunataks (lonely peaks). They peaked out above the glacial ice and are thought to have remained ice-free during glacial periods. Several remote coastal islands such as Coronation Island located off the southern Panhandle may have also remained ice-free during glacial periods. These places served as refugia for plants and animals during glacial periods. Recent research in these unique areas is helping scientists better understand species that may have predated the last glacial period and be quite ancient compared to other species in Southeast Alaska.

Several proposed rare and endangered plant species have been found on Haida Gwaii just south of Alaska. No doubt with additional botanical study, similar new species may be found in Southeast Alaska as well.

Not covered in this guide are invasive species, species introduced into an area where they did not evolve and, therefore, have no natural enemies to impede their spread. *Invasive Plants of Alaska*, a cooperative endeavor by several Federal and State agencies and programs, as well as the University of Alaska, Fairbanks, is an excellent reference for these species.

It is my hope that this field guide will contribute more pieces to the complex and fascinating puzzle that is today's native flora of Alaska.

KEY TO FAMILIES

KEY TO GROUPS

1a. Plants with microphylls (small, scale-like leaves, usually with a single vein); reproduction by means of spores; flowers or woody cones lacking. **(2)**
1b. Plants with macrophylls (large leaves, usually with more than a single vein); reproduction by spores or seeds, the latter borne in flowers or cones. **(3)**

2a. Stems not jointed; leaves green and partly overlapping, not whorled or forming a sheath at the node; plants either aquatic and grasslike or terrestrial and resembling mosses or miniature conifers; sporangia borne in axils of leaves or modified leaves in cones.
CLUB MOSSES AND RELATIVES
2b. Stems jointed and fluted; leaves not green, reduced to a whorl of connate scales at the nodes; plants neither grass- nor moss- nor coniferlike; sporangia borne under shield- shaped, scalelike structures in terminal cones. **HORSETAILS (FAMILY EQUISETACEAE)**

3a. Plants fernlike, with broad leaves, reproducing by spores; flowers and cones lacking. **FERNS**
3b. Plants generally not fernlike, reproducing by seeds; flowers or cones present. **(4)**

4a. Seeds borne naked on the surface of a scale, these crowded together on an axis and forming a cone; flowers lacking.
CONIFERS
4b. Seeds borne in ripening carpels; plants with flowers. (FLOWERING PLANTS) **(5)**

5a. Cotyledons 2; stems mostly increasing in diameter by means of a cambium between the xylem and phloem; leaves mostly net-veined; flower parts usually in 4's or 5's. **DICOTYLEDONEAE**
5b. Cotyledons 1; stems usually lacking a cambium, or if a cambium is present, it produces entire vascular bundles; leaves mostly parallel-veined; flower parts usually in 3's. **MONOCOTYLEDONEAE**

KEY TO CLUB MOSSES AND RELATED FAMILIES

1a. Plants aquatic, submersed in ponds or lakes, or occasionally growing on exposed mud; leaves grasslike, long and slender, from broadly clasping bases; sporangia borne embedded in the leaf base. **ISOETACEAE**
1b. Plants terrestrial, growing in various habitats, but usually not in ponds or lakes; leaves scalelike or awl-shaped; sporangia borne in terminal strobili or at leaf bases along the stem. **(2)**

2a. Plants with spores of one kind; sterile leaves lacking a ligule; strobili circular in transverse section.
LYCOPODIACEAE (*Lycopodium*)
2b. Plants producing two kinds of spores (male and female); sterile leaves with a ligule at the base; strobili quadrangular (circular in transverse section in *S. selaginoides*). **SELAGINELLACEAE (*Selaginella*)**

KEY TO FAMILIES OF FERNS

1a. Sporangia borne naked, on erect spike- or paniclelike segment of normal frond arising from base of blade or from petiole.
OPHIOGLOSSACEAE
1b. Sporangia borne on margin or underside of the frond, usually with fertile and sterile portions of blade similar, dimorphic in some. **(2)**

2a. Sori marginal, enclosed by a 2-valved margin of frond; leaf blade
membranous, one cell thick; sporangia with oblique annulus.
HYMENOPHYLLACEAE
2b. Sori on the underside of the frond or almost marginal; leaf
blade several cells thick; sporangia with a vertical annulus.
POLYPODIACEAE

KEY TO FAMILIES OF CONIFERS

1a. Leaves alternate or almost opposite, 2-ranked, seed cones reduced to
a single seed with a fleshy, finally red-colored, cup-shaped aril; plants
dioecious. **TAXACEAE**
1b. Leaves opposite; whorled or spirally arranged,not 2-ranked; seed cones
well developed, woody or fleshy but not as above; plants monoecious
or dioecious. **(2)**

2a. Shrubs or trees with opposite or whorled, scalelike or linear-awlshaped
leaves; cones mostly less than 15 mm long, the scales opposite or
whorled, woody or fleshy. **CUPRESSACEAE**
2b. Mostly trees with needles spirally arranged or borne in clusters; cones
mostly more than 15 mm long, the scales spirally arranged, woody to
papery. **PINACEAE**

KEY TO DICOTYLEDONEAE KEYS

1a. Plant a tree or woody ascending shrub. **KEY I**
1b. Plant not a tree, if shrub then mature plant prostrate or dwarfed and without large
woody stems. **(2)**

2a. Flowers with sepals and petals fused or sepals and petals lacking. **KEY II**
2b. Flowers with sepals and petals distinct. **(3)**

3a. Petals separate. **KEY III**
3b. Petals fused, at least from base. **KEY IV**

DICOT KEY I. FAMILIES OF TREES AND TALL SHRUBS

1a. Plants parasitic on branches of trees; rooting in the host (*Tsuga*), usually yellow green. **LORANTHACEAE (*Arceuthobium*)**
1b. Plants not parasitic on branches of trees, rooting in the soil. **(2)**

2a. Leaves opposite. **(3)**
2b. Leaves alternate. **(7)**

    3a. Leaves and branches with starlike hairs; perianth in a single whorl; fruit a red drupe. **ELEAGNACEAE (*Shepherdia*)**
    3b. Leaves and branchlets hairless or hairy, but not with starlike hairs; perianth of sepals and petals, or if sepals only, then the fruit not a drupe. **(4)**

    4a. Leaves palmately veined and lobed or parted; fruit a double samara. **ACERACEAE (*Acer*)**
    4b. Leaves pinnately veined, simple or compound; fruit drupaceous or capsular. **(5)**

    5a. Ovary superior; stamens 2-5; fruit a capsule; plants decumbent subshrubs; stamens 5; fruit a 2-3 valved capsule. **ERICACEAE**
    5b. Ovary inferior; stamens 4-5; fruit fleshy, non splitting, one seeded. **(6)**

    6a. Petals united; stamens 5 (rarely 4). **CAPRIFOLIACEAE**
    6b. Petals separate; stamens mostly 4. **CORNACEAE (*Cornus*)**

7a. Perianth consisting of a single whorl (arbitrarily called sepals), or lacking (see also CHENOPODIACEAE). **(8)**
7b. Perianth consisting of both sepals and petals. **(13)**

    8a. Plants evergreen, trailing; leaves with narrow rolled blades. **EMPETRACEAE (*Empetrum*)**
    8b. Plants deciduous; leaves mostly with expanded blades. **(9)**

    9a. Leaves and branchlets with starlike hairs; flowers solitary or in axillary clusters. **ELAEAGNACEAE (*Elaeagnus*)**
    9b. Leaves and branches variously hairy or hairless, but not with starlike hairs; flowers in catkins. **(10)**

    10a. Buds with a single visible scale; dioecious subshrubs, shrubs, or trees. **SALICACEAE (*Salix*)**
    10b. Buds with few to several visible scales; dioecious or monoecious shrubs or trees. **(11)**

    11a. Leaves ovate, once-toothed; dioecious trees; fruit a 2-4-valved capsule. **SALICACEAE (*Populus*)**
    11b. Leaves various in shape, but if ovate to lanceolate then doubly-toothed, or else the plants monoecious shrubs; fruit a nutlet or waxy drupe. **(12)**

12a. Leaves narrowly oblanceolate to flattened-spoon-
shaped, toothed only near the apex; fruit a waxy drupe.
MYRICACEAE (*Myrica*)
12b. Leaves ovate to lanceolate or orbicular, toothed
from near the base; fruit a nutlet, this often winged.
BETULACEAE

13a. Corolla of united petals. **(14)**
13b. Corolla of separate petals. **(16)**

14a. Flowers in dense heads, a cluster of bracts subtending the flower heads; leaves mostly lobed or divided.
ASTERACEAE (*Artemisia*)
14b. Flowers variously displayed, but without a cluster of bracts subtending the flower heads; leaves mostly not lobed or divided. **(15)**

15a. Stigma 3-lobed; ovary 3-loculed; fruit a 3-valved capsule.
DIAPENSIACEAE (*Diapensia*)
15b. Stigma capitate; ovary 2-10-loculed; fruit a capsule, berry, or drupe.
ERICACEAE

16a. Stamens numerous, more than twice as many as the petals.   ROSACEAE
16b. Stamens few, not more than twice as many as the petals. **(17)**

17a. Plants depressed shrubs or subshrubs; leaves simple, pinnately veined, very
small. **(18)**
17b. Plants erect or ascending shrubs; leaves palmately veined, often lobed, moderate to large. **(19)**

18a. Sepals 3, petal-like; petals 3; stamens 3; fruit a fleshy,
black berry; leaves deeply grooved beneath, congested.
EMPETRACEAE (*Empetrum*)
18b. Sepals 4; petals 4, reflexed; stamens 8; fruit a red berry; leaves flat,
not congested on the stem.   ERICACEAE (*Vaccinium oxycoccos*)

19a. Leaves with spines along the petioles and main veins; stems coarse and spiny;
inflorescence terminal.   ARALIACEAE (*Oplopanax*)
19b. Leaves lacking spines along the petioles and main veins; stems slender, armed
or unarmed; inflorescence lateral.   GROSSULACEAE (*Ribes*)

KEY II FLORAL ENVELOPE WITH THE COROLLA AND CALYX COMBINED, OR LACKING

1a. Plants parasitic on branches of trees, rooting in the host (*Tsuga*), usually yellowish green. **LORANTHACEAE (*Arceuthobium*)**
1b. Plants not parasitic on branches of trees, rooting in soil. **(2)**

2a. Leaves alternate or basal and alternate. **(3)**
2b. Leaves opposite or whorled, at least on the lower portion of the stem (see also CHENOPODIACEAE). **(12)**

    3a. Leaves odd-pinnately compound or palmately lobed, the lobes with sharp forward pointing teeth on the margin; stipules present, not sheathing. **ROSACEAE (*Sanguisorba*)**
    3b. Leaves various, but not as above; stipules absent, or if present then sheathing. **(4)**

    4a. Flowers sessile, in involucrate heads; anthers united into a tube around the style. **ASTERACEAE**
    4b. Flowers variously arranged, but usually not in heads; anthers not joined. **(5)**

    5a. Flowers perigynous, 4-merous; stamens 4 or 8, inserted in the notches of a disk which almost fills the center of the flower; leaves orbicular, with 6-10 broadly rounded teeth. **SAXIFRAGACEAE (*Chrysosplenium*)**
    5b. Flowers epigynous, perigynous, or hypogynous, usually not 4-merous; stamens not inserted on a disk (except in Apiaceae and Santalaceae); leaves not as above. **(6)**

    6a. Ovary inferior; plants with perfect flowers. **(7)**
    6b. Ovary superior or plants dioecious. **(8)**

        7a. Leaves simple, well distributed along the stem; flowers in axillary or terminal cymes; fruit fleshy, one- seeded, non splitting. **SANTALACEAE (*Comandra*)**
        7b. Leaves mostly once or thrice compound; flowers in compound umbels; styles 2; fruit dry, splitting into two halves. **APIACEAE**

8a. Pistils several to many per flower; stamens usually 10-many. **RANUNCULACEAE**
8b. Pistils one per flower; stamen usually 10 or less. **(9)**

9a. Plants aquatic, usually more or less submerged; perianth lacking. **CALLITRICHACEAE (*Callitriche*)**
9b. Plants terrestrial, sometimes growing in moist soil; perianth present. **(10)**

10a. Ovary 2-loculed; stamens 2, 4, or 6; fruit a silicle. **BRASSICACEAE (*Lepidium*)**
10b. Ovary 1-loculed; stamens 2-9; fruit an achene or utricle. **(11)**

11a. Plants with a stipular sheath (ocrea) above each node; perianth segments mostly 6; fruit an achene. **POLYGONA-CEAE**

11b. Plants lacking stipules; leaves often mealy with collapsed globular hairs; flowers inconspicuous, sepals 1-5; corolla lacking; fruit a utricle. **CHENOPODIA-CEAE**

12a. Plants aquatic; leaves simple, or if dissected then ovary 4-loculed; stigmas 4. **HALORAGA-CEAE**

12b. Plants terrestrial; leaves various, but not as above. **(13)**

13a. Flowers sessile, in dense heads; anthers united into a tube around the style. **ASTERA-CEAE**

13b. Flowers not in dense heads; anthers not united. **(14)**

14a. Leaves pinnately parted or compound; inflorescence of terminal or axillary, many- flowered cymes. **VALERIANACEAE (*Valeriana*)**

14b. Leaves palmately lobed or parted, or simple and neither lobed nor parted; inflorescence various. **(15)**

15a. Leaves palmately lobed or parted; stem leaves in 1-2 whorls of 3. **RANUNCULACEAE (*Anemone*)**

15b. Leaves simple, not lobed or parted; stem leaves several or lacking, not in whorls of 3. **(16)**

16a. Plants with stinging hairs; flowers inconspicuous, in axillary spikes. **URTICACEAE (*Urtica*)**

16b. Plants lacking stinging hairs; flowers not in axillary spikes. **(17)**

17a. Flowers in terminal umbels, subtended by 4 large petal-like bracts (the whole resembling a single, terminal flower). **CORNACEAE (*Cornus*)**

17b. Flowers variously arranged, but not subtended by petal-like bracts. **(18)**

18a. Ovary inferior, 2-loculed; flowers in cymes. **RUBIACEAE (*Galium*)**

18b. Ovary superior, 1-loculed; flowers axillary, or in racemes. **(19)**

19a. Styles 2-5; calyx segments distinct. **CARYOPHYLLACEAE**

19b. Styles 1; calyx segments united. **PRIMULACEAE (*Glaux*)**

KEY III. COROLLA WITH SEPARATE PETALS

1a. Stamens more than twice as many as the petals. **(2)**
1b. Stamens few, not more than twice as many as the petals. **(5)**

> **2a**. Plants aquatic; leaves immersed or floating, heart- or shieldshaped; carpels united or separate (in *Brasenia*). **NYMPHAEACEAE**
> **2b**. Plants terrestrial, or if aquatic and with separate carpels, the leaves not shield- shaped. **(3)**
>
> **3a**. Ovary inferior, or partly so. **ROSACEAE**
> **3b**. Ovary superior. **(4)**
>
> **4a**. Stamens, petals, and sepals all attached to the margin of a hypanthi- um; flowers perigynous. **ROSACEAE**
> **4b**. Stamens, petals, and sepals all attached to the base of the ovary; flow- ers hypogynous. **RANUNCULACEAE**

5a. Pistils 3 or more. **(6)**
5b. Pistils 1 (rarely 2). **(9)**

> **6a**. Plants succulent, fleshy; sepals, petals, and pistils 4-5 each. **CRASSULACEAE (*Sedum*)**
> **6b**. Plants not succulent; pistils often more numerous than the sepals and petals. **(7)**
>
> **7a**. Plants aquatic; leaves shield-shaped. **NYMPHAEACEAE (*Brasenia*)**
> **7b**. Plants terrestrial, or if aquatic, then the leaves not shield-shaped. **(8)**
>
> **8a**. Stamens inserted on a hypanthium; flowers perigynous. **ROSACEAE**
> **8b**. Stamens inserted at the base of the ovary; flowers hypogynous. **RANUNCULACEAE**

9a. Ovary inferior, or partly so. **(10)**
9b. Ovary superior. **(14)**

> **10a**. Plants aquatic; leaves mostly whorled, simple or finely dissected; stamens 1 or 4-8. **HALORAGACEAE**
> **10b**. Plants terrestrial, sometimes growing in wet places. **(11)**
>
> **11a**. Inflorescence a compound umbel; leaves mostly ternate or finely divided; styles 2; fruit dry, splitting into two halves, or a berry. **APIACEAE (*Umbelliferae*)**
> **11b**. Inflorescence a raceme or panicle, or a simple umbel (or the flowers solitary); fruit and leaves not as above. **(12)**
>
> **12a**. Inflorescence a simple umbel subtended by 4 large petal-like bracts. **CORNACEAE (*Cornus*)**
> **12b**. Inflorescence in a branched (paniculate) or unbranched raceme (or the flowers solitary), not subtended by petal-like bracts. **(13)**
>
> **13a**. Corolla of 2 or 4 petals; stamens 2 or 8. **ONAGRACEAE**
> **13b**. Corolla of 5 petals; stamens 5 or 10. **SAXIFRAGACEAE**

14a. Styles 2-5 (appearing 6-10 in **DROSERACEAE**), distinct to near the base. **(15)**
14b. Styles 1, sometimes lobed or divided at the apex, or style obsolete. **(18)**

15a. Leaves with long, glandular, sensitive hairs which trap insects.
DROSERACEAE (Drosera)
15b. Leaves variously pubescent or hairless, but not as above. **(16)**

    16a. Leaves opposite.   CARYOPHYLLACEAE
    16b. Leaves alternate, or basal and alternate, or if opposite (as in some Portulacaceae), then the sepals 2. **(17)**

    17a. Sepals 2; stamens opposite the petals or fewer (5-3).
PORTULACACEAE
    17b. Sepals 5; stamens alternate with the petals or more numerous (except 3 in Tolmiea).   SAXIFRAGACEAE

18a. Sepals 2-3. **(19)**
18b. Sepals 4, 5 or more. **(21)**

19a. Corolla regular.   PORTULACACEAE
19b. Corolla irregular. **(20)**

    20a. Sepals 3, with 2 small and green, the third petal-like and produced backward into a recurved spur.   BALSAMINACEAE (Impatiens)
    20b. Sepals 2, equal or nearly so, lacking a spur (the corolla sometimes spurred).   FUMARIACEAE (Corydalis)
21a. Flowers irregular. **(22)**
21b. Flowers regular. **(24)**

    22a. Flowers pea-like, the upper petal (banner) enclosing the others in bud; fruit a legume; leaves mostly compound.
FABACEAE (Leguminosae)
    22b. Flowers not pea-like; fruit a capsule; leaves simple. **(23)**

    23a. Flowers solitary, the petals violet to yellow or white.
VIOLACEAE (Viola)
    23b. Flowers few to numerous, the petals pink, greenish yellow, or creamy white.   PYROLACEAE (Pyrola)

24a. Sepals and petals 4; stamens 6 (4 plus 2), 4, or rarely 2.
BRASSICACEAE (Cruciferae)
24b. Sepals and petals (4)5(7); stamens (4)5-10. **(25)**

25a. Leaves palmately lobed, cleft, or divided, with stipules; carpels tailed at maturity; separating from each other as 1-seeded, indehiscent segments.
GERANIACEAE (Geranium)
25b. Leaves simple pinnate; lacking stipules; carpels not as above. **(26)**

26a. Stamens 5, alternating with broad, dissected, glandular-tipped staminodia; ovary 1-loculed.   SAXIFRAGACEAE (Parnassia)
26b. Stamens 5-more; staminodia lacking; ovary more than 1-loculed.
PYROLACEAE

KEY IV. COROLLA OF PETALS UNITED AT LEAST NEAR THE BASE.

1a. Ovary inferior, or partly so. **(2)**
1b. Ovary superior (see *Menyanthaceae*). **(7)**

2a. Stamens more than 5; anthers opening by terminal pores.
ERICACEAE
2b. Stamens 5 or less; anthers not opening by terminal pores. **(3)**

3a. Stamens united by their anthers; flowers sessile in involucrate heads; stamens adnate to the corolla. **ASTERACEAE**
3b. Stamens separate. **(4)**

4a. Leaves alternate; flowers mostly large and more or less bellshaped.
CAMPANULACEAE (*Campanula*)
4b. Leaves opposite or whorled; flowers mostly small and not bellshaped. **(5)**

5a. Leaves pinnately lobed or compound; stamens 2- 3(4); flowers irregular; plants herbaceous; frequently ill scented. **VALERIANACEAE (*Valeriana*)**
5b. Leaves simple, or pinnately compound; stamens 4-5; flowers regular or irregular; plants herbaceous or woody; mostly not ill scented. **(6)**

6a. Plants woody shrubs, vines, or trailing subshrubs; leaves opposite (rarely whorled) or perfoliate; fruit 1-several seeded. **CAPRIFOLIACEAE**
6b. Plants herbaceous; leaves opposite or whorled; fruit 2-seeded.
RUBIACEAE (*Galium*)

7a. Stamens more than 5 (mostly twice as many as the corolla lobes). **(8)**
7b. Stamens 5 or less (if more, then the same number as the corolla lobes). **(11)**

8a. Corolla segments distinctly united, saucer, cup, or tube shaped. **(9)**
8b. Corolla segments not markedly united, usually connate only near the base, or only part of the segments united. **(10)**

9a. Anthers opening by terminal pores. **ERICACEAE**
9b. Anthers opening by a longitudinal slit. **PYROLACEAE**

10a. Petals 5; fruit a legume or loment.
FABACEAE
10b. Petals 4, in 2 unlike pairs; fruit a capsule with 2 parietal placentae.
FUMARIACEAE (*Corydalis*)
11a. Plants parasitic, lacking chlorophyll; leaves small, scale-like. **OROBANCHACEAE**
11b. Plants usually not parasitic, with chlorophyll; leaves usually well developed. **(12)**

12a. Corolla irregular. **(13)**
12b. Corolla regular. **(16)**

13a. Ovary with 1 ovule per locule, appearing 4-loculed, 4-lobed; fruit of 4 nonsplitting, 1-seeded nutlets; stems mostly square; plant often with a mintlike odor. **LAMIACEAE**
13b. Ovary with more than 1 ovule per locule, usually not 4-loculed nor

4-lobed; stems mostly round; plants lacking a distinctive odor. **(14)**

14a. Plants aquatic; leaves often dissected and bearing small bladders; corolla spurred; stamens 2. **LENTIBULARIACEAE (*Utricularia*)**

14b. Plants usually terrestrial, or if aquatic then not as above. **(15)**

15a. Plants with a scape bearing a single, blue to purple flower. **LENTIBULARIACEAE (*Pinguicula*)**

15b. Plants without a scape; inflorescence few- to many-flowered. **SCROPHULARIACEAE**

16a. Sepals 2; styles 3, or 3-lobed; stamens 2-5. **PORTULACACEAE**

16b. Sepals (4)5 or more, or connate with (4)5 or more lobes; styles various; stamens various. **(17)**

17a. Stamens as many as the corolla lobes and opposite them. **PRIMULACEAE**

17b. Stamens as many as the corolla lobes and alternate with them, or fewer. **(18)**

18a. Corolla small (2 mm broad or less), thin, dry, membranous, veinless; capsule opening by a lid; leaves mostly basal; inflorescence a dense spike. **PLANTAGINACEAE (*Plantago*)**

18b. Corolla various, but not as above; leaves mostly along the stem; inflorescence mostly other than a spike. **(19)**

19a. Ovary 4-lobed, appearing 4-loculed; fruit consisting of 4 nutlets at maturity. **BORAGINACEAE**

19b. Ovary not 4-lobed, the locules 1-3; fruit not consisting of nutlets. **(20)**

20a. Style 3-cleft; ovary 3-loculed, or the stigma 3-lobed; fruit a 3-valved capsule. **(21)**

20b. Style not 3-cleft; ovary 1- to 2-loculed; fruit not as above. **(22)**

21a. Plants dwarf perennials with single- flowered inflorescence borne on a naked or single-bracted peduncle and usually with 3 bracts at the base of the calyx; stigma merely 3-lobed. **DIAPENSIACEAE (*Diapensia*)**

21b. Plants various, but if dwarf perennials, the inflorescence not as above; styles 3-cleft. **POLEMONIACEAE**

22a. Ovary 1-loculed. **(23)**

22b. Ovary with 2 or more locules. **(25)**

23a. Leaves mostly basal, from a stout rhizome, trifoliolate, or broadly orbicular to kidney-shaped; scapes bearing compound cymes or racemes. **MENYANTHACEAE**

23b. Leaves along the stem, at least some neither trifoliolate nor broadly orbicular, or if orbicular, then with few, large teeth. **(24)**

24a. Leaves opposite, entire; styles 1 or none. **GENTIANACEAE**

24b. Leaves alternate or basal and alternate; or if opposite, then pinnatifid; styles 2 or 2-cleft apically. **HYDROPHYLLACEAE**

25a. Plants aquatic; leaves strictly basal, long- petioled, the blades elliptic to spatula-shaped; margins entire. SCROPHULARIACEAE (*Limosella*)

25b. Plants terrestrial, sometimes growing in wet soil; leaves not strictly basal or if long- petioled then margins not entire. **(26)**

26a. Styles 2 or 2-cleft apically; fruit a capsule. HYDROPHYLLACEAE (*Phacelia*)

26b. Styles 1; stigma entire or merely lobed apically; fruit a capsule or berry. **(27)**

27a. Principal leaves basal, long petiolate, the blades orbicular to kidney-shaped, palmately veined with few large teeth or lobes.
HYDROPHYLLACEAE (*Romanzoffia*)

27b. Principal leaves along the stem; mostly short petiolate or sessile, the blades mostly ovate to lanceolate; pinnately veined, entire to sharply toothed or pinnately lobed; stamens 2; flowers pale blue to blue. SCROPHULARIACEAE (*Veronica*)

KEY TO FAMILIES OF MONOCOTYLEDONAE

1a. Flowers borne on a spadix (a fleshy spike) subtended by a spathe (an enlarged, often showy bract). ARACEAE

1b. Flowers variously arranged, but not on a spadix; spathelike bracts lacking, or if present, then 2-more and seldom showy. **(2)**

2a. Perianth well-developed, at least the inner segments petal-like in color and texture. **(3)**

2b. Perianth lacking or reduced and inconspicuous, its parts often of bristles or scales, not petal- like in texture or color. **(5)**

3a. Ovary superior; stamens 6 or 4. LILIACEAE

3b. Ovary inferior; stamens 3, 2, or 1. **(4)**

4a. Fertile stamens 3; flowers regular or nearly so. IRIDACEAE

4b. Fertile stamens 1-2; flowers irregular. ORCHIDACEAE

5a. Flowers sessile in the axils of chaffy or husklike scales; leaves with sheathing bases; perianth usually much reduced. **(6)**

5b. Flowers not both sessile and in the axils of chaffy bracts; leaves often without sheathing bases, or if sheathing basally, then perianth well developed. **(7)**

6a. Leaves 2-ranked on the stem; stems mostly hollow and circular in transverse section; anthers capable of turning, attached crosswise on the apex of the filament; flowers subtended by 2 bracts (palea and lemma).
POACEAE (Graminae)

6b. Leaves 3-ranked; stems mostly solid, triangular in transverse section (or sometimes almost circular in transverse section); anthers attached or fixed by the base; flowers subtended by a single bract.
CYPERACEAE

7a. Plants terrestrial, or growing in shallow water, usually leaves and flowers both emergent. **(8)**

7b. Plants floating or submerged aquatics, usually not raised above the surface of the water. **(11)**

    8a. Inflorescence a spike or spike-like elongate raceme.
                                     JUNCAGINACEAE
    8b. Inflorescence of almost globular heads or racemes or otherwise, but not in spikes. **(9)**

    9a. Flowers imperfect, the lower heads pistillate; perianth of chaffy scales.
                            SPARGANIACEAE (*Sparganium*)
    9b. Flowers perfect, usually not in heads (although they may appear head-like in some species); perianth 6-parted, in 2 whorls. **(10)**

    10a. Pistils 3 per flower, divergent; fruit a follicle.
                         JUNCAGINACEAE (*Scheuchzeria*)
    10b. Pistils 1 per flower; fruit a capsule.   **JUNCACEAE**

11a. Plants strictly maritime; flowers in flattened spikes.   **ZOSTERACEAE**
11b. Plants not maritime (although sometimes growing in brackish water); flowers variously arranged, but not in flattened spikes. **(12)**

12a. Flowers in few-rayed umbels; leaves seldom over 1 mm wide.
                                 RUPPIACEAE (*Ruppia*)
12b. Flowers in spikes or heads; leaves mostly over 2 mm wide. **(13)**

13a. Flowers perfect, usually in elongate spikes; pistils 4 per flower.
                     POTAMOGETONACEAE (*Potamogeton*)
13b. Flowers monoecious, in heads (pistillate below, staminate above); pistils 1 per flower.   **SPARGANIACEAE** (*Sparganium*)

CLUB MOSSES AND QUILLWORTS

ISOETACEAE
QUILLWORT FAMILY

Isoetes echinospora   Spiny-spored Quillwort

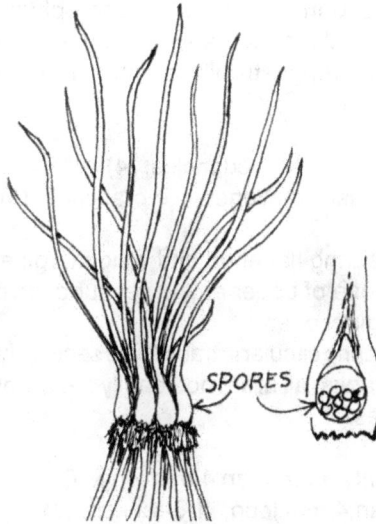

SPORES

Aquatic, grass-like evergreen perennial about 10 cm tall with spirally arranged fronds in basal rosettes. Fronds hollow, flaccid, often lying flat on the pond or lake bottom, tapering to a long, fine-pointed tip, 2.5- 5 cm long. Stem thick, fleshy, corm-like, flattened on top. Few stomata on leaf tips. Sporangia oblong to oval at base of frond on inner side. Spores of two types: powdery microspores tiny and numerous, smooth; megaspores few to 1/2 mm in diameter when dry, chalky white, with three radial ridges, microscopically spiny.

Southeast Alaska north to the Arctic Circle in shallow, slightly acidic ponds and lakes on sand or gravel. Uncommon or perhaps overlooked. Quillworts produce spores borne in sporangia in their leaf axils. Also known as *I. muricata, I. braunii, I. setosa* and *I. maritima.*

LYCOPODIACEAE
CLUBMOSS FAMILY

KEY TO SPECIES OF *LYCOPODIUM:*

1a. Sporophylls appearing like vegetative leaves, elongate, several times longer than wide. **(2)**

1b. Sporophylls distinctly different from vegetative leaves, commonly 1-2 times longer than wide. **(3)**

    2a. Sporangia borne in axils of ordinary foliage leaves, not forming strobili; stems perennial and evergreen, arising in small, erect clusters: plants widespread, frequently bearing bud-like gemmae among leaves.   *L. selago*

    2b. Sporangia borne in axils of sporophylls aggregated into terminal sessile cone; stems annual, prostrate with erect branches bearing strobili; plants of extreme southeastern Alaska.   *L. inundatum*

3a. Strobili borne on slender peduncles. **(4)**

3b. Strobili borne sessile on vegetative branches. **(5)**

    4a. Stems appearing flattened and wing margined; leaves 4-ranked, those of upper and lower surfaces differing from decurrent marginal ones.   *L. complanatum*

    4b. Stems appearing circular in transverse section, not margined; leaves in about 10 ranks, all alike, not readily separable in upper or lower stem surfaces.   *L. clavatum*

5a. Leaves commonly more than 4 mm long, 6-10-ranked. **(6)**

5b. Leaves less than 4 mm long, 4-5-ranked. **(7)**

    6a. Prostrate stems subterranean; erect stems much branched, resembling a small conifer; leaves mostly less than 5 mm long; plants restricted to southern Alaska.   *L. dendroideum*

    6b. Prostrate stems above ground; erect stems simple or with few long, erect, ascending branches, not especially conifer-like; leaves mostly more than 5 mm long; plants of broad distribution.   *L. annotinum*

7a. Vegetative branches flattened; leaves 4-ranked, decurrent maginal ones unlike those of upper and lower leaves.   *L. alpinum*

7b. Vegetative branches circular in transverse section; leaves (4)5-ranked, all alike.   *L. sitchensis*

Lycopodium alpinum   Alpine Clubmoss

Stem creeping, elongate, branches erect, 2-10 cm tall, paired, constantly forking, hairless, tufted. Leaves bluish green, 4-ranked, mostly in two forms: on lower surface trowel-shaped, others somewhat flattened. Mostly stalkless sporangia in strobili 1-2 cm long at ends of branched leafy stems; sporophylls yellow, broadly triangular, abruptly contracted at base.

Most of Alaska south of the Brooks Range. Alpine tundra, bogs and subalpine forests, occasional in lowland forest. Infrequent. *L. sitchense* has round stems and light green leaves alike and 5-ranked. *L. complanatum* in contrast has longer fruiting body stalks and awl-shaped, flat lying leaves, occurring in low elevation forests.

Lycopodium annotinum   Stiff Clubmoss

Horizontal, above ground stem with erect, simple or paired aerial stems to 25 cm tall, annual growth conspicuous. Leaves 8-ranked, 4 leaves in each whorl, all similar, reflexed or divergent to ascending with short, stiff point. Single cones stalkless and solitary at ends of leafy branches; sporophylls ovate with toothed margins, straw-colored; sporangia kidney-shaped.

Most of Alaska south of the Brooks Range. Wet woods, bogs and dry tundra. Common forest clubmoss. Spores used as a dusting powder for condoms, as powder to stop bleeding, increase urine flow and to treat diarrhea. Chewing stems produces slight intoxication, mouth pain, vomiting and diarrhea. *L. clavatum* in contrast has stalked cones and creeping stems.

Lycopodium clavatum   Running Clubmoss

Stem long, creeping, densely covered with mainly flat lying hairs, ascending branches much forked with age. Leaves bright green, mostly 10-ranked, lying flat, all similar, ends twisted and tipped with soft white hair-like brushes. Spores borne in 1-3 cylindric cones at end of long, erect stalk, yellowish, sporophylls ovate with white, thread-like tips, sporangia kidney-shaped.

Most of Alaska south of the Brooks Range. Forest, dry open areas and subalpine areas. Common. Used medicinally for liver and gallbladder problems and digestive disorders. *L. annotinum* in contrast has spreading leaves, solitary stalkless cones and longer side branches.

Lycopodium complanatum   Ground-cedar

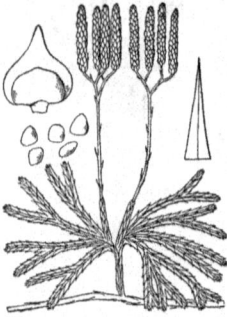

Stem extensively creeping, elongate, branches erect, cedar-like to 35 cm tall, forked, branchlets flattened. Leaves somewhat whitish-green, 4-ranked, united over half their length, ventral leaves smaller and awl-shaped, lying flat against stem. Spores in cylindric cones in clusters of 2 on long stalks; sporophylls yellowish, round to triangular; sporangia kidney-shaped.

Much of Alaska south of the Brooks Range except the Aleutians; in southeastern Alaska found in Lynn Canal area. Low elevation, dry forests. Locally common. Our only clubmoss with cedar-like flattened sprays. Also known as *L. subinaefolium.* *L. alpinum* in contrast has smaller, shorter fruiting body stalks in tighter clusters and trowel shaped leaves.

Lycopodium dendroideum   Ground-pine

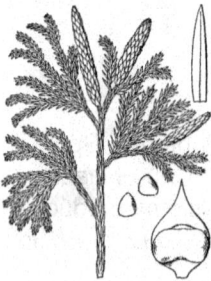

Scattered, upright, tree-like branches to 30 cm tall from deeply subterranean rhizomes. Leaves numerous, 6 to 9-ranked, prickly, long, linear, tapering and spreading. Cones cylindric, stalkless, solitary on upper branches.

Southeastern Alaska west to Aleutians, scattered in interior central Alaska. Dry woodlands and second-growth shrubby areas. Also known as *L. obscurum*.

Lycopodium inundatum   *Bog Clubmoss*

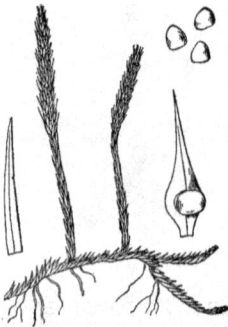

Short creeping stems 5-10 cm long, simple or rarely branched. Leaves spiral, 8- or 10-ranked, strongly upturned, soft yellowish green color, similar on both fruiting and non fruiting stems, often only one fertile stem. Spores in single erect stalkless cones; sporophylls green, similar to foliage leaves, sporangia almost spherical.

Dixon Entrance north to Baranof Island. May be more widespread. Bogs and wet lake shores at low elevation. Locally common. Unique among clubmosses in having a stalkless, terminal cone.

Lycopodium selago   Fir Clubmoss

Horizontal rooting stems short and withering, erect stems tufted, simple or branched in pairs with all branches reaching same height. Leaves 8-ranked, lance-shaped, alike, almost lying flat or wide spreading, usually hollow at base, often with dilated gemmae in axils.

Throughout Alaska. Low elevation to subalpine forests, alpine tundra, open areas and high elevation or snowfield-influenced bogs. Common. Unique among the clubmosses in having spore-bearing bodies at the base of leaves rather than in strobili and in not having a creeping stem. Contains poisonous alkaloid that causes mouth pain, vomiting and diarrhea. May contain a substance effective against Alzheimer's disease. The spores of all clubmosses are flammable and used in fireworks and early flash photography. Now classified as **Huperzia selago**.

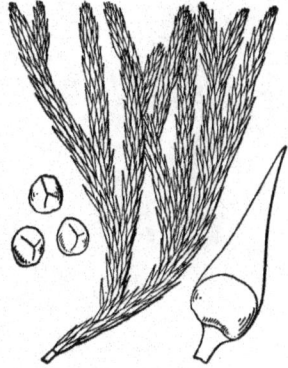

Lycopodium sitchense   Sitka Clubmoss

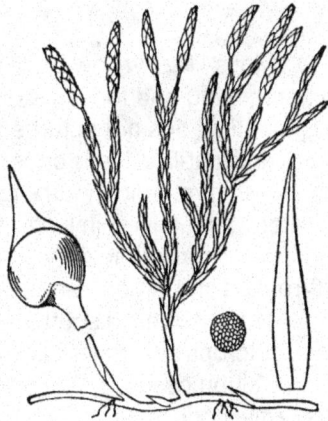

Horizontal rooting stems short and withering, erect stems tufted, simple or branched in pairs with all branches reaching same height. Leaves 8-ranked, lance-shaped, alike, almost lying flat or wide spreading, usually hollow at base, often with dilated gemmae in axils. Horizontal elongate stem to 50 cm or more, rooting at intervals, numerous erect stems 7-15 cm tall, branching dichotomously several times, forming compact tufts, sterile branches not flattened. Leaves awl-shaped, strongly tapering, mostly 5-ranked. Solitary cones at ends of short stems with oppressed, scale-like leaves, margins rough to the touch, appearing gnawed.

Coasts and islands of southern Alaska. Alpine meadows and woods, rock outcrops. Common. May be growing with **L. alpinum** which is 4-ranked without trowel-shaped leaves and with flattened branches.

SELAGINELLACEAE
LESSER-CLUBMOSS FAMILY

Selaginella selaginoides Mountain Spikemoss

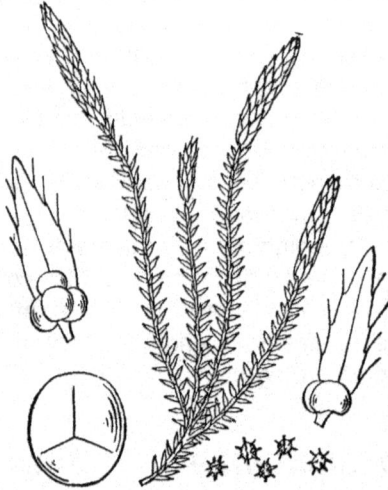

Moss-like, yellow green perennial forming loose to dense mats with creeping and forking stems on rock or ground. Stems not readily fragmenting, tips not upturned, consisting of two types: sterile stems weak and mat-like, fertile ones thread-like and ascending. Leaves without dorsal groove, spirally arranged, uniform, lance-like to elliptic, sharp pointed, minutely spiny on margins. Cones inconspicuous, slender, almost cylindrical, solitary, 2-4 cm long, stalkless; sporophylls mostly 10-ranked and larger than leaves, more prominently spined and nerved.

Much of Alaska south of the Brooks Range. Muskegs, wet banks, rock outcrops and wet meadows, in neutral to alkaline soil. Occasional. Unlike clubmosses, spikemosses have inconspicuous cones because the fertile cone leaves only slightly differ from sterile leaves and spores are borne in sporangia in modified, sterile leaf axils. This species is thought to be a primitive member of the genus. It is unique in having a dispersal mechanism, termed "compression and slingshot megaspore ejection". Another species, *S. siberica,* may occur in the Lynn Canal area.

6

SPHENOPSIDA: THE HORSETAILS

EQUISETACEAE
THE HORSETAILS

KEY TO SPECIES OF *EQUISETUM*:

1a. Stems evergreen, usually unbranched or if branched, then lacking regular whorls of branches; cones with peduncles seldom exceeding subtending sheath with minute pointed tip. **(2)**

1b. Stems annual, not evergreen, often consisting of 2 types: some usually with regular whorls of branches; cones with at least some peduncles much surpassing subtending sheath, rounded apically. **(3)**

    2a. Stems slender, 5-10(12)-ridged, commonly 10-40 cm tall; teeth of leaf sheaths persistent, central cavity to 1/3 stem diameter.   *E. variegatum*

    2b. Stems stout, (14)16-20-ridged or more, commonly 20-100 cm tall or more; teeth of sheaths jointed, deciduous; central cavity taking up 3/4 or more of stem.   *E. hyemale*

3a. Plants bearing strobili in summer; fertile and sterile stems alike, green, ridges of stem smooth or minutely cross-wrinkled, lacking small tubercles or spicules. **(4)**

3b. Plants bearing strobili in spring (less commonly in summer); fertile and sterile stems not alike; ridges of stems with small tubercles or spicules or almost smooth, but not cross-wrinkled. **(5)**

    4a. Stems deeply 5-10-grooved, central cavity about 1/6 diameter of stem; teeth of leaf sheaths (2.5) 3-7 mm long, conspicuously white-margined; central cavity less than 1/3 diameter of main stem.   *E. palustre*

    4b. Stems shallowly 9-25-grooved; teeth of leaf sheaths 1.5-3.5 mm long, not or scarcely white-margined; central cavity more than 1/2 diameter of main stem.   *E. fluviatile*

5a. Fertile stems whitish, pinkish, yellowish, or brownish, soon withering; ridges of stem with minute bumps or cross-ridges; branches of sterile stems not again branched.   *E. arvense*

5b. Fertile stems becoming green and branched, persistent; ridges of stem with small tubercles, small spicules or transverse ridges; branches of sterile stems branched.   *E. pratense*

7

Equisetum arvense　　Common Horsetail

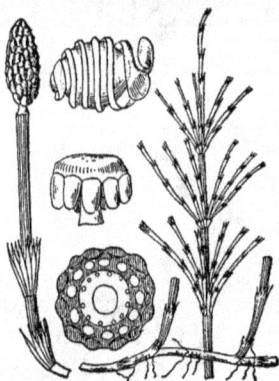

Rhizome with soft fine hairs and ovoid tubercles. Annual shoots to 70 cm tall of two types: sterile stems solitary or clustered, 10-12 ridges with minute bumps, cross ridges and branches in regular whorls; fertile stems whitish to brownish, borne in springtime, soon whithering; sterile sheaths with persistent teeth, separate or sometimes united, brown or blackish, fertile sheaths with teeth; stomates on sterile fronds in two broad bands, not sunken, apically blunt. Cones long-stalked, blunt tipped, persistent.

Throughout Alaska. Arctic and alpine tundra, woods, beaches, lake margins and disturbed sites. Common. Succulent fertile shoots eaten when fresh. Variable species with numerous forms. *E. pratense*, which it resembles, has green, persistent, branched fertile stems, first internode of primary branches on sterile stem shorter than or equal to stem sheath which has white-tipped teeth. Cell walls of all horsetails consist of silicon dioxide, used as an abrasive for cleaning and polishing. Rhizome used by Tlingit people to decorate baskets. Horsetails as well as ferns grew to small tree size when dinosaurs roamed the planet.

Equisetum fluviatile　　Swamp Horsetail

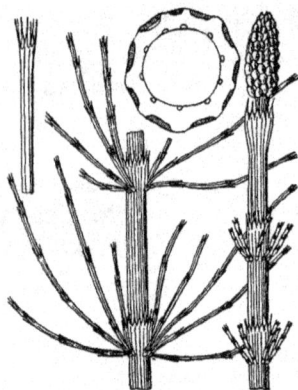

Rhizome hairless, shiny, reddish, rarely with tubercles, fertile and sterile stems alike, to one meter tall. Large, annual stems which collapse when squeezed, simple or branched with wide central cavity and fine groves, 10-30 ridged, branches single or in irregular whorls at internodes, large diameter central cavity, branching mostly above the middle; sheaths green with dark-brown teeth; stomates in one band in each groove, not sunken. Spores borne in cones at tip of main stem, short-stalked, blunt-tipped, deciduous.

Much of Alaska, except northern coastal regions and Aleutians. Shallow water, marshy places at low elevation. Common, often occurring in great numbers. Forms hybrids with *E. arvense*. Reported to cause poisoning in horses in south-central Alaska.

Equisetum hyemale Common Scouring-rush

Rhizome with dense soft hairs, growing deep in soil. Stems perennial to 50 cm tall, evergreen, fertile and sterile stems alike, unbranched, thick, strongly grooved; sheaths cylindrical, usually with 2 black bands separated by grayish band at maturity; stomates in two rows in each grove, not sunken. Spores borne in short-stalked cone with minute pointed tip.

Southeastern Alaska west to Aleutians. Sandy shores, roadsides and wet open areas. Occasional, uncommon north of the 50th parallel. Used for polishing and sharpening; also used in China to cure conjunctivitis, influenza and colds, dysentery and edema. *E. variegatum* in contrast is less than 50 cm tall and less than 4 mm thick.

Equisetum palustre Marsh Horsetail

Rhizome hairless, thin, dark reddish brown, lacks soft fine hairs or tubercles. Stems annual, to 60 cm tall, smaller central stem cavity, deeply grooved, all similar with 5-10 deep ridges, simple or usually with single to several thick branches of differing lengths, branches usually lacking near top of stem; first sheath of branches very short, dark brown, next green, shorter than corresponding sheath of stem; stomates in broad band in each grove, not sunken. Spores in cones on top of main stem, long-stalked, obtuse, soon withering.

Glacier Bay to south-central Alaska and interior Alaska. Shallow water, wet marshy places and roadsides. Common. *E. fluviatile* in contrast has a larger central stem cavity and more stem ridges.

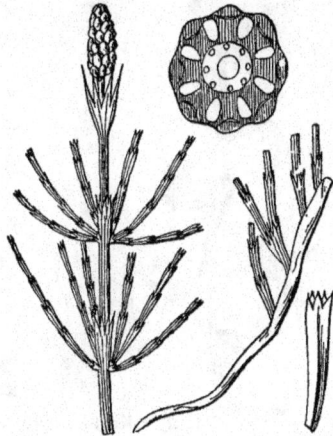

9

Equisetum pratense   Meadow Horsetail

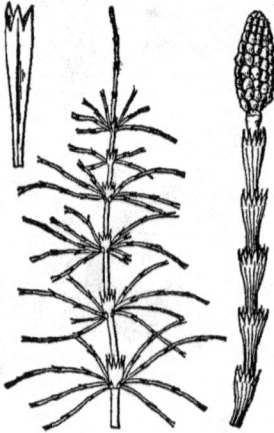

Rhizome black and creeping. Stems scattered, 16-50 cm tall, annual, of two types: sterile stems pale green, mostly solitary, fertile stems unbranched at first, later developing whorls of branches similar to those on the sterile stems; sheaths greenish-white, almost cylindrical with persistent teeth, sheaths on the fertile stems more conspicuous and somewhat longer; stomates in two bands in grooves, not sunken. Spores in obtuse cones, long-peduncled, soon withering.

Most of Alaska except northern coastal plain. Meadows, wet woodlands, beaches and grassy stream banks. Occasional in southeastern Alaska; common further north.

Equisetum variegatum   Northern Scouring-rush

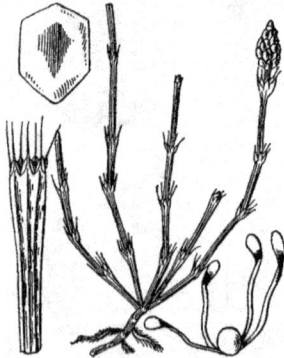

Rhizome black and shiny. Stems perennial, clustered, evergreen to 50 cm tall, fertile and sterile stems alike with 5-12 ridges less than 4 mm thick, ridges finely 1-grooved with 2 rows of tubercles; teeth all black or with obscure white margins, 8--14, incurved, sheaths flared, base not easily distinguished, black or blackish tips, stomates in 2 rows in each groove, sunken below skin. Cone almost stalkless or short stalked with small predominately pointed tip, maturing in late summer, or cones overwintering and shedding spores in spring.

Throughout Alaska. Muskegs, stream banks, lake shores, wet woods and tundra. A variable species with several ecotypes, some of which are distinct subspecies. Hybrids of this species with *E. hyemale* reported.

PTERIDOPHYTA: THE FERNS

HYMENOPHYLLACEAE
FILMY FERN FAMILY

Hymenophyllum wrightii   Filmy Fern

Rhizome thread-like. Plant small and moss like, branched, forming loose mats. Fronds lax and delicate, hairless, triangular-ovate, 2-3 pinnatifid, 2-3 by 1-1.5 cm, compound, stipes thread-like, blackish, hairless except for small tuft of hair-like scales at base. Sori solitary in cup- or purse-like openings at tips of veins; indusium of ovate valves, spores four-sided.

Coasts and islands of southeastern Alaska. On wet vertical cliffs and rocks, epiphytic on bark and decaying wood of conifers or among mosses and liverworts. Rare or perhaps overlooked. Although sporophytes of this species are known from a single site on Queen Charlotte Islands, British Columbia, gametophytes are more common along the coasts of Alaska and B.C. The gametophytes reproduce vegetatively and can persist and spread via gemmae without needing a sporophyte generation. Also known as *Mecodium wrightii.*

OPHIOGLOSSACEAE
ADDER'S-TONGUE FAMILY

Botrychium lunaria   Common Moonwort

Fleshy, hairless 4-30 cm tall. Sterile blade somewhat leathery, pinnate, pinnae in 2-5 pairs with entire or wavy margins, rarely toothed, stalkless or on short stalk, broadly fan-shaped, often overlapping except in shaded forest forms, blade oblong in outline, fertile portion longer and branched, venation appears like the ribs of a fan, midribs absent; indusium lacking, sporangia borne on panicle-like clusters and lacking annulus, opening by a regular split, spores numerous and four sided.

Most of Alaska south of the Brooks Range. Grassy slopes, sandy soils, bogs and meadows. This species grows with other species of moonwort, occasionally hybridizing with them and is the most widespread of the moonworts.

11

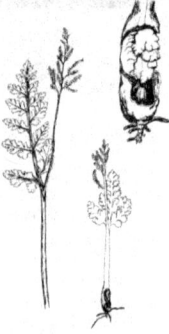

Botrychium boreale (Northern Grapefern) Deciduous, 4-25 cm tall, stout and fleshy, egg-shaped, stalkless blade with broader oblong segments, fertile spike longer than the sterile leaf, triangular sterile blade borne high on stem and pinnae with distinct median vein. Can be found growing with **B. lunaria** in alpine meadows, stream banks, bogs and grassy slopes. Coasts and islands of southern Alaska. Middle to subalpine elevation. Rare. Also known as ***B. pinnatum.***

Botrychium lanceolatum (Lance-leaved Grapefern) Deciduous, 5-20 cm tall, stalkless, sterile triangular blade borne high on the stem, with narrow lance-shaped segments attached below the much branched fertile spike, stout and fleshy, stalk of fertile segment much shorter than sterile blade. Scattered populations in alpine meadows, open forests, sandy clearings of southern Alaska. Middle to alpine elevation.

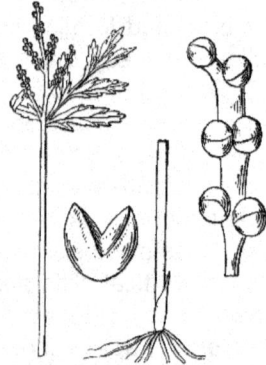

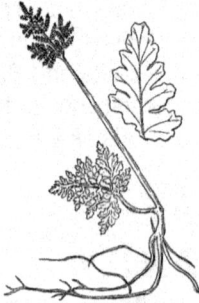

Botrychium multifidum (Leathery Grapefern) Erect, 10-50 cm tall, thick, leathery, triangular sterile blade attached near stem base below ground, sterile frond of previous year may persist until September, fertile frond on stalk as tall or taller than sterile leaf. Open woods, meadows and bogs, coasts and islands of southern Alaska. Possibly a hybrid between **B. lunaria** and **B. lanceolatum**. Also known as ***B. matricariae.***

Botrychium virginianum (Rattlesnake Grapefern) Erect, 25-75 cm tall. Blades to 20 cm wide, deciduous, glabrous, broadly triangular, much dissected, thin rather than fleshy or leathery, nearly stalkless, borne above middle of stem. Open grassy slopes, wet meadows, deciduous forests, coasts and islands of southern Alaska. The common name refers to fertile leaf resembling tip of a rattlesnake. Grapeferns are not true ferns since their sporangia are borne in grape-like clusters on a naked stalk, rather than on leaves as in true ferns.

POLYPODIACEAE
COMMON FERN FAMILY

KEY TO GENERA OF *POLYPODIACEAE*:

1a. Fronds conspicuously dimorphic, fertile with contracted pinnae. **(2)**

1b. Fronds not dimorphic, fertile and vegetative, much alike. **(3)**

    2a. Plants with both fertile and vegetative fronds once pinnately compound, pinnae of fertile leaves rolled in from edges, entire with continuous narrow sori; plants of coasts and islands of southern Alaska.   ***Blechnum***

    2b. Plants with both fertile and vegetative fronds more than once pinnately compound or with pinnae pinnatifid with sori various, but not as above; plants of various distribution.   ***Cryptogramma***

3a. Sori borne along margin of lower surface of frond, covered by enrolled indusium-like margin. **(4)**

3b. Sori borne on lower surface of frond, not covered by margin, naked or with true indusia. **(5)**

4a. Fronds scattered and twice to three times pinnately compound with main central axis; sori continuous along margin of reflexed frond segment; blades and rhizomes hairy, not scaly; stipe thick, tough, yellow to brown.   ***Pteridium***

4b. Fronds fan-shaped and twice compound with dichotomously branching rachis, sori clearly separate, borne on underside of reflexed margins of pinnae, blades hairless, stipe and rhizome scaly, not hairy; stipe slender, brittle, black.   ***Adiantum***

5a. Indusia lacking, even on young fronds, sori naked. **(6)**

5b. Indusia present, at least on young fronds, sori more or less protected. **(9)**

    6a. Fronds once pinnately compound or pinnatifid, pinnae entire or merely toothed, not again pinnatifid; evergreen, petioles jointed to rhizome.   ***Polypodium***

    6b. Fronds twice to three times pinnately compound or ternately compound or pinnatifid, pinnae lobed or again compound, herbaceous; petioles continuous with rhizome. **(7)**

    7a. Fronds at least three times pinnately compound, densely crowded on short, stout rhizome, lanceolate in outline.   ***Athyrium***

    7b. Fronds twice pinnately compound or pinnae merely pinnatifid or if three times pinnately or ternate-pinnately compound, then broadly triangular in outline. **(8)**

    8a. Fronds more or less ternately compound, pinnae once to twice pinnately compound or pinnatifid, blade broadly triangular, often broader than long, hairless; pinnae jointed to rachis.   ***Gymnocarpium***

    8b. Fronds pinnately compound, pinnae pinnatifid, blade ovate to triangular in outline, longer than broad, more or less hairy at least when young, lower pinnae not jointed to rachis.   ***Thelypteris***

9a. Indusium elongate, flap-like, attached along sorus on side next to edge of leaf segment, sometimes curved apically but not kidney-shaped. **(10)**

9b. Indusium shield-shaped, kidney-shaped, or attached below sorus, not as above. **(11)**

    10a. Blades small, commonly less than 25 cm long; fronds once pinnately compound, evergreen, with toothed pinnae about as broad as long.   ***Asplenium***

    10b. Blades large, commonly more than 25 cm long; fronds twice to three times pinnately compound or pinnatifid, pinnae several times longer than broad.   ***Athyrium***

11a. Fronds evergreen, more or less leathery, often sharply toothed; indusium shield-shaped, attached at center and opening all around.   ***Polystichum***

11b. Fronds deciduous or less commonly evergreen, not leathery or sharply toothed; indusium not shield-shaped. **(12)**

12a. Indusium orbicular to kidney-shaped, attached at sorus; plants mostly course with conspicuous scaly petioles; fronds often more than 25 cm long (except in *Dryopteris fragrans*). **(13)**

12b. Indusium hood-like or plate-like, borne beneath sorus; plants mostly delicate and slender with scaly or smooth and hairless petioles; fronds commonly less than 25 cm long. **(14)**

    13a. Fronds once pinnately compound, pinnae deeply pinnatifid, pubescent along upper surface and usually along side of rachis with unicellular hairs; 2 vascular bundles in petiole.   *Thelypteris*

    13b. Fronds twice to three times pinnately compound, merely scaly or minutely glandular along rachis; 3 or more vascular bundles in petiole.   *Dryopteris*

14a. Indusium hood-like, attached at one side, partly under sorus and commonly pushed back as sorus expands; veins reaching leaf margin; fronds not accompanied by conspicuous, persistent petiole bases.   *Cystopteris*

14b. Indusium plate-like, borne symmetrically beneath sorus, splitting into radiating segments; veins not reaching leaf margin; fronds accompanied by conspicuous, persistent, petiole bases.   *Woodsia*

Caution: Many species of ferns produce fiddleheads that are poisonous especially if under cooked. Several Alaskan species are suspected of containing the enzyme thiaminase which can break down vitamin B1 unless cooked. If you must eat them, cook well before eating and consume in moderate amounts. Collecting more than half the fiddleheads from a single rhizome, year after year, will result in lower yields and eventually kill the fern.

Adiantum pedatum   Maidenhair Fern

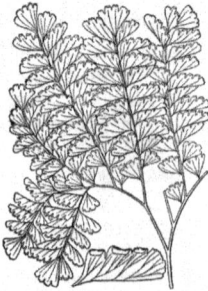

Rhizome slender and dark-colored. Fronds erect to 100 cm tall, fan-shaped, 2-pinnate, delicate, deciduous, grayish-green, resistant to water drops, stipes erect, shiny black or deep reddish brown, wiry, up to 60 cm long, each tiny leaf division fringed along upper edge, leaf blades soft and thin, fronds form concentric circles. Sori marginal, not continuous, on edges of upper lobes of leaflets, covered by indusia formed by reflexed tips of lobes. Indusium transversely linear or linear-oblong.

Coast and islands from southeastern Alaska to Aleutians. Wet areas, wet rock cliffs, stream banks, in the spray zone of waterfalls, often on humus-covered talus slopes and moist limestone soils. Occasional to common. Fronds used in tea for treating coughs, hoarseness and breathing difficulties. Thin leaves good for drying berries. Also known as *A. aleuticum*.

Asplenium viride   Green Spleenwort

Rhizome short with blackish scales. Densely tufted to 14 cm tall, stipes with deep groove, brown in lower part becoming greenish upwards, rachis delicate, green or greenish. Fronds numerous, lax and ascending, densely tufted, pinnae not evergreen, blades narrow, linear or linear-lance-shaped. Sori elongated in 2-4 pairs in sausage-like lines along indistinct mid vein; indusium membranous, conspicuous and attached along one side of rachis, entire.

Coasts and islands from southeastern Alaska to Aleutians. Seeps, talus slopes, wet rocky places, especially on limestone or basic rocks. Middle to alpine elevation. Occasional. Small, easily overlooked fern. Species name "*viride*" refers to green rachis of frond. In Old English, the word "wort" meant plant. Europeans once believed that when a plant resembled a human organ, it cured diseases of that organ, in this case, the spleen.

Asplenium trichomanes (Maidenhair Spleenwort) Generally larger than *A. viride*, leaflets firm, evergreen and mostly opposite, stipe bare, dark chestnut brown to purplish-black, rachis colored like stipe, stipe and rachis persist for several years and may be more numerous than living fronds. Rock outcrops, crevices and talus slopes on islands in central southeastern Alaska where evidently rare. Found at lower elevation than *A. viride*.

Athyrium filix-femina   Common Lady Fern

Rhizome stout with small membranous scales. Plant large and graceful with sweeping curves like a giant parentheses, fronds widest near middle and tapering evenly toward top and bottom, to 2 m tall, stems straw-colored and scaly, pinnae linear to lance-shaped-oblong, pinnules thin with scalloped margins. Sori straight, hooked at distal end or horseshoe-shaped; covered by oblong shield when young; indusium straight or curved, often toothed.

Southeastern Alaska west to Aleutians. Open moist places and moist woods especially along streams. Common. A variable and wide-ranging species. Can be distinguished from *Dryopteris expansa* by its tapering fronds. Fiddleheads good boiled, baked or steamed. Rhizomes edible if steamed overnight. See caution on pg. 14.

Athryrium distentifolium (Alpine Lady Fern) Rhizome very stout covered with brown scales, fronds shorter than those of ***A. filix-femina***, 20-80 cm long, appearing scorched and wrinkled, pinnae narrowly triangular, pinnules thin, stalked, oblong-lance-shaped to narrowly triangular. Sori numerous, round, less than 1 mm wide; indusium rudimentary or lacking. Moist alpine thickets in coastal and island areas of southeastern and south-central Alaska. Also known as ***A. alpestre, A. americanum*** and ***Phegopteris alpestris.***

Blechnum spicant   Deer Fern

Rhizome short and woody with narrow chestnut colored scales, to 60 cm tall, evergreen, in close tuft or rosette. Fronds of two types: fertile fronds longer with pinnae commonly narrower and longer with very narrow leaflets, vegetative fronds gradually narrow toward base, leathery, forming circular crown. Sori along margins of segments covering entire lower surface; indusium conspicuous, brownish, membranous.

Southeastern Alaska west to Kenai Peninsula and eastern Kodiak Island, rarely on Aleutians. Forests and forest edge, subalpine meadows and along streams. Leaves used as a medicine for skin sores. Important food for deer.

Cryptogramma crispa   Parsley Fern

Rhizome short, compactly branched, clothed with scales, old leaf bases and tufted leaves to 20 cm tall. Fronds numerous, densely clustered, of two types: fertile fronds linear and longer than vegetative fronds, vegetative fronds straw-colored to greenish, ovate to lance-like stipes, blades 2-3 times pinnately compound. Sori more or less continuous, covered by folded margins of ultimate segments.

Most of Alaska south of the 65th parallel. Rock crevices and rocky slopes. Occasional, more common at high elevation. Also known as ***C. acrostichoides;*** includes ***C. sitchensis.***

16

Cystopteris fragilis   Fragile Fern

Rhizome light brown, creeping, unbranched, densely scaly. Delicate fronds to 30 cm tall, usually clustered, stipes straw-colored, hairless, scape scaly brownish at base. Blade lance-shaped to elliptic, 2-3 times pinnately compound, hairless or nearly so, pinnae once or twice pinnately compound, ultimate segments toothed or lobed. Sori round, borne along veins on lower side of blade; indusium attached under sorus, delicate, hood-like, arching over sorus, often pushed back as sorus enlarges, soon withering.

Much of Alaska except northern coastal plain. Open rocky, often calcium-rich forests, cliffs and wet rocks at all elevations. Occasional. Extremely variable with respect to leaf dissection and spore sculpture.

Dryopteris expansa   Spiny Shield-fern

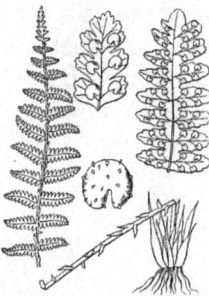

Rhizome stout, ascending, covered with old stipe bases. Fronds deciduous, 30 to 100 cm tall, finely cut and lacy, bare leaf stalk folded under, whole leaf triangular, lowest pair of leaflets largest and turned at different angles than upper leaflets. Sori rounded, brownish, covered by a rounded indusium when young.

Southeastern Alaska west to Aleutians. Forests and forest edges, along streams. Common. Rhizomes gathered in spring and fall, fiddleheads gathered before fronds loosen and steamed or boiled and peeled; roots used to make dye, tea for kidney trouble and asthma. Songbirds use fine roots to line nests. One of our most common and widespread ferns. Also known as *D. dilatata, D. assimilis* and *D. austriaca*. See caution on pg. 14.

Dryopteris fragrans (Fragrant Shield-fern) Smaller than *D. expansa*, less than 13 cm tall, densely clustered evergreen leaves, sickly sweet aroma when fresh and crushed, scaly fronds, old leaves persistent as gray or brown conspicuous clump at plant base, lowermost pinnae smaller than those near middle of blade. Sori midway between midvein and margin of segments; indusium glandular. Talus slopes and rock outcrops from Juneau north throughout Alaska except far northern parts.

17

Gymnocarpium dryopteris   Oak Fern

Rhizome creeping and slender. Stipe shiny, black and hairless, 20-40 cm tall. Fronds deciduous, usually solitary or in masses, thin and lace-like, scattered, blade triangular, divided into three once or twice again pinnately compound segments, terminal segment larger than or equal to lateral segments. Sori roundish, borne along veins on lower side of blade; indusium lacking.

Much of Alaska south of the 65th parallel. Forests, along streams and seeps. Very common, forming continuous carpet in forest understory. Also known as *Phegopteris dryopteris,* includes *G. disjunctum*. Closely related to genus *Dryopteris. Gymnocarpium,* Greek for gumnos (gymnos), "naked" and karpos (karpos), "fruit"; a reference to the lack of indusia; dryopteris, from the Greek, drus (drys), "oak", pteris (pteris), "fern".

Polypodium glycyrrhiza   Licorice Fern

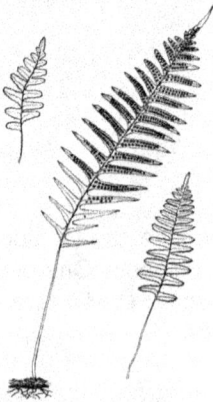

Rhizome sometimes with licorice taste, creeping, densely covered by reddish-brown uniformly colored scales. Evergreen, petioles straw-colored to greenish, lacking scales. Fronds to 70 cm tall, single, greenish gray to coarse, segments sparsely pubescent with gray or brownish hairs and smooth margins, pinnae generally in 10 to 20 offset pairs, toothed, pointed. Sori rather large, oval to round on rows parallel to midrib of leaflet; indusium lacking.

Southeastern Alaska west to Aleutians. Beach forests, rocks, epiphyte on tree trunks and branches. Common. Rhizomes contain same chemical that gives licorice its flavor: ostadin, a steroidal compound 3000 times sweeter than sucrose. Rhizomes used raw or roasted as cough, sore throat and cold medicine. Also known as *P. vulgare*. The name *Polypodium* means "many feet", referring to its branched rhizomes; *glycyrrhiza* translates as "sweet root".

Polystichum braunii   Braun's Holly Fern

Rhizome, erect, stout and densely covered with scales and old leaf bases to 1 m tall. Petioles brown to straw-colored, course, scaly. Fronds numerous, blade elliptic to lance-elliptic, twice pinnately compound, pinnules with bristle-tipped teeth. Sori in two rows near mid-vein; indusium shield-shaped, often irregularly toothed or eroded as if gnawed.

Southeastern Alaska west to Aleutians. Moist forests, streamsides and rocky slopes. Occasional. Some taxonomists recognize a separate species *P. andersonii* distinguished by presence of one or more buds in axils of upper pinnae. Includes *P. setigerum*. Hybrids with *P. lonchitis* have been reported from southeastern Alaska.

Polystichum lonchitis   Mountain Holly Fern

Rhizome densely covered with old leaf bases and scales, leaves tufted, simply pinnate. Fronds ascending in close crown to 60 cm tall, short stipes, densely chaffy with lustrous, rusty-brown scales, blades hard, rather stiff, evergreen, leathery. Sori large, touching in 2 single or double rows; indusia large, irregularly toothed.

Southeastern Alaska west to Aleutians. Dry coniferous forests, more commonly on cliffs and talus slopes from mid to subalpine elevation. Occasional. Fronds narrower than those of *P. braunii.*

Polystichum munitum   Swordfern

Rhizome stout, woody, ascending, densely covered with reddish-brown scales and old leaf bases. Stipes densely chaffy. Fronds stiffly erect to 1.5 m tall, blades evergreen, lance-elliptic to oblong, leathery, once pinnately compound, petioles course, scaly throughout, pinnae alternate, often scaly beneath, sharply 2-toothed. Sori in one to several rows; indusium round with fringed margins.

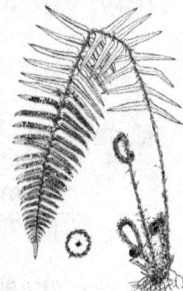

Southeastern Alaska south of Frederick Sound. Well-drained forests. Common. Rhizome good roasted. See caution on pg. 14. Used traditionally as flooring and bedding and to line pit ovens.

Pteridium aquilinum   Bracken Fern

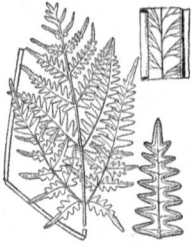

Rhizome branched, pubescent, buried deep in soil, to 1 m tall, coarse appearance, tending to form a dense cover, stems stout with dark woolly mass basally, straw-colored above. Fronds erect, thick, course and fibrous, leaf blade triangular, usually 3 times pinnately compound, basal pair of pinnae often larger and more strongly dissected. Sori marginal, protected by downward bending margin of leaflet; indusium apparently lacking.

Southeastern Alaska except Yakutat area. Dry open places, muskeg-forest fringe, lake shores and meadows. Occasional; rare to nonexistent north of Juneau. Fiddleheads should not be eaten because they are reported to cause stomach cancer as well as livestock poisoning. Mature leaves contain enzyme thiaminase which is poisonous. This is the world's most widespread fern species.

Thelypteris phegopteris   Northern Beech Fern

Rhizome long, slender, hairy, not covered with old leaf bases. Stipe hairy and scaly, pale straw-colored. Single frond deciduous to 40 cm tall, petiole longer than blade, hairy and scaly, blade triangular to egg-shaped, once or twice pinnately compound, pinnae hairy on both surfaces, lower most pinnae largest and usually drooping in mature plants. Sori small, round, borne near margin; indusium lacking.

Much of Alaska south of the 65th parallel except central and eastern Alaska. Moist open forests along streams, stony slopes and meadows to lower alpine. Common. Also known as *Phegopteris connectilis.*

Thelypteris limbosperma   Mtn. Wood Fern

Rhizome short, stout, covered with persistent leaf bases. Fronds tufted, elliptic to oblanceolate, yellow-green leaf blade, lower surface with numerous brownish-yellow glands, lower most pinnae much smaller than those at middle of blade, hairy on upper side of rachis, strongly lemon-scented. Sori round, borne near margins; indusium kidney- or horseshoe-shaped, falling early.

Southeastern Alaska west to Aleutians. Woods, meadows, stream banks and alpine tundra in acid soil. Occasional in southeastern, common in south-central Alaska. Also known as *T. oreopteris* and *T. quelpaertensis*.

Woodsia ilvensis   **Rusty Cliff Fern**

Rhizome erect, compact, with abundant persistent petiole bases of more or less equal length, scales uniformly brown, lance-like, fronds to 20 cm long, deciduous, in dense tufts, leaf segments considerable longer than broad with scales and hairs below; stipe jointed, stout, dark brownish yellow below, paler above, scaly and hairy; pinnae 10 to 20 pair, lowest somewhat reduced, sessile or nearly so, 4-9 pairs of pinnules, mostly opposite. Sori numerous, round near the margin; indusium with numerous long hair-like segments, conspicuously overtopping sporangia.

North of Juneau to Brooks Range. Dry rocks and scree slopes. Also known as ***Acrostichum ilvense.***

Woodsia glabella (Smooth Cliff Fern) Rhizome short, stipe with joint from base, about 5 cm tall, light-colored, lacking scales or hairs, blade yellowish-green. Sori few, but frequently coming together: indusium deeply cleft into a few delicate, narrow hair-like segments that overtop the sporangia. Throughout Alaska. Calcareous rocks. Uncommon in southeastern Alaska but found in Misty Fiords and in the alpine at Quartz Hill near Ketchikan.

Woodsia scopulina (Rocky Mountain Cliff Fern) Rhizome stoutish, short, creeping; fronds numerous in large tufts to 30 cm tall, stipes not jointed, with soft downy hair, blades linear to oblong-lanceolate with soft, flat, whitish hairs mixed with glands; sori above middle; indusium of flat, scale-like segments, more or less concealed sporangia. Open slopes mostly on limestone soils, dry crevices and talus slopes in coastal northern southeast and south-central Alaska.

21

GYMNOSPERMAE
THE CONIFERS

CUPRESSACEAE
CYPRESS FAMILY

Chamaecyparis nootkatensis Alaska Yellow-Cedar

Tree 12 to 25 m tall (shrub at higher elevation or in peat bogs) spreading and drooping branches, twigs 4-angled or somewhat angled, bark gray or purplish, scaly, peeling in long narrow strips, yellowish inner bark with distinctive raw potato-like odor. Leaves small, scale-like in four rows, pointed and spreading, closely overlapping, bluish-green. Cones short-stalked, nearly round, hard, ashy gray; 4 or 6 paired, rounded, hard cone scales. Seeds 2-4 under a cone-scale.

Southeastern Alaska west to Wells Bay in Prince William Sound; alpine meadows, bogs and transitional bog-forests. Locally abundant around Sitka and Peril Strait on Baranof and Chichagof islands, scattered in the rest of Southeast Alaska. Cones ripen 2nd year. Wood used to make canoe paddles, dishes, spoons, adze handles and masks. Also used for utility poles, furniture, window frames, doors, boats, etc. Fine roots used for baskets and hats. Wood decay resistant. Oldest tree species in Southeast Alaska and can exceed 1000 years of age Seeds difficult to germinate; reproduces mainly by layering. Alaska Yellow-cedar is an economically and culturally valuable tree. Since the end of the Little Ice Age in the late nineteenth century, it has experienced an increasingly accelerated rate of mortality resulting in large stands of dead trees in largely pristine Southeast Alaskan and Canadian forests. Studies suggest decreasing snow accumulations at lower elevations due to warming climate, may allow small surface roots to die, since they are no longer being insulated by snow.

Juniperus communis   Common Juniper

Low or prostrate spreading evergreen shrub forming mats or clumps, bark gray or dark reddish brown, rough scaly and thin. Leaves in whorls of 3, spreading at right angles or curved slightly downwards, awl-shaped, sharp pointed, usually with a white band on upper surface. Cones lateral on very short scaly stalks, berry-like, blue, covered with a whitish bloom, hard, mealy, resinous and sweet, maturing to black the second or third season. Seeds 3 or fewer, light brown, pointed.

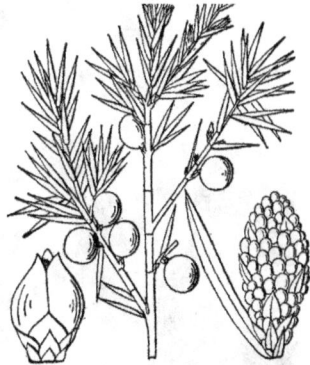

Southeastern Alaska west to Prince William Sound and Kenai Peninsula, Cook Inlet and Interior Alaska. Dry slopes and rock outcrops in alpine or subalpine and bogs. Scattered. Berries edible and used as seasoning for meat dishes. Needles burned on woodstoves as incense. Boiled berries and leaves used for colds. Most widely distributed conifer in the world and most widespread tree species in north temperate zone.

Thuja plicata   Western Redcedar

Large evergreen tree to 30 m tall and 6-12 cm in diameter with swollen base, pointed conical crown, leafy twigs flattened in fan-like sprays, slightly drooping, bark gray or brown, fibrous, becoming thick and furrowed into long ridges. Needles scale-like, flattened, pointed, shiny yellow green above. Cones clustered near ends of twigs, elliptic, light brown, composed of several paired elliptic leathery scales. Seeds three or fewer under a cone-scale with 2 narrow wings.

Southeastern Alaska north to Port Frederick Sound near Petersburg. In boggy to moderately well drained soils in forests and bogs from sea level to 300 m in elevation and can exceed 1000 years of age. Abundant. Wood used for totem poles and dugout canoes; stringy bark used for mats, baskets, ropes and clothing; produces a smokeless fire good for drying fish; used as a drill for starting friction fires. A "slow match" made by binding a bundle of shredded cedar bark tightly with some less flammable substance. When inner core is ignited, it smolders for hours, ready to ignite at any time; used as torches by native people.

23

PINACEAE
PINE FAMILY

Abies amabilis Pacific Silver Fir

To 25 m tall or more, up to 1 m in diameter, twigs finely hairy, bark smooth, gray, splotched with white. Needles crowded and spreading, lacking stomates above, two broad silvery bands of stomates below, flattened, tending to form flattened sprays. Cones in highest branches, oblong, upright, purplish, staminate flowers red. Seeds light brown with broad wing.

Small scattered populations in extreme southeastern Alaska: east and south of Ketchikan on well drained soils away from salt spray to 300 m in elevation. Grows with western hemlock and Sitka spruce. Uncommon but locally abundant. Cones ripen in autumn. Pitch used on canoe paddles and other wooden articles; boughs used as bedding and floor coverings. Needles pulverized and used as baby powder and body scent.

Abies lasiocarpa Subalpine Fir

Small or medium-sized tree with long narrow crown, growing prostrate and shrubby at high elevation. Also forms carpet-like growth as little as 15 cm tall near timberline. Bark thin, gray, smooth except for numerous resin blisters on young trees, becoming shallowly fissured; young twigs reddish hairy. Needles crowded, spreading and curved upward by a twist at base, flattened, blunt or notched, deep blue-green, stomates on all sides. Cones in highest branches, oblong, upright, dark purple, finely hairy, with fan-shaped scales, staminate flowers bluish. Seeds light brown with broad wing.

Scattered populations throughout southeastern Alaska. Largest populations in Misty Fjords and in a pure stand northeast of Skagway, in areas of recent glaciation, such as valley bottoms, or on moist subalpine slopes near timberline. Reported but not verified in Copper River basin and Prince William Sound. Isolated, relict populations on high mountains on northern Prince of Wales and central Dall Islands. Forms subalpine forest with mountain and western hemlock and Sitka spruce, or nearly pure stands on recently deglaciated outwash plains. Locally abundant. Soft, even-grained wood used to make roofing shingles; pitch makes a good teeth cleaner when chewed.

Picea sitchensis   Sitka Spruce

Tree 45-70 m tall and 1-3 m in diameter from greatly enlarged base; tall, straight trunk, bark gray and smooth becoming dark purplish brown with scaly plates. Needles stand out on all sides of twig, flattened and slightly keeled, sharp with two whitish bands of stomates on upper surface. Cones short-stalked and hanging down, cone scales long, stiff, rounded and irregularly toothed. Seeds brown with large wing.

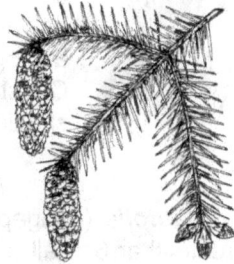

Southeastern Alaska west to Kodiak Island and north coastal Alaska Peninsula. Along streams, rocky slopes, below cliffs, on shorelines, sea level to alpine. Pure stands occur following deglaciation, fire, landslides, or along exposed shorelines and in large river floodplains. Abundant. Often propagates by layering; young shoots used for tea, leaf buds for making spruce-tip beer, roots for baskets, hats, ropes, gum for chewing; wood light and durable and highly prized for airplanes (Spruce Goose) boats and sailboat masts. Hybrids with *P. glauca* occur west of Prince William Sound near its limit of range in south-central Alaska. Lives 500 to 750 years. This is Alaska's state tree.

Pinus contorta   Shore Pine

Low spreading or scrubby tree 2-12 m tall and 25-40 cm in diameter, 23 m tall and 55-65 cm in diameter on well-drained, sunny sites; bark brownish to blackish, scaly, thin or becoming thick and furrowed. Needles two in a bundle with sheath at base relatively long and stiff, often twisted, yellow green to dark green. Cones one or few, almost stalkless, egg-shaped, pointing backward on twig, opening at maturity, persisting on tree for many years and opening irregularly. Seeds brown.

Southeastern Alaska north to Yakutat. Bogs, sandy glacial outwash, forest edges on benches near lakes from sea level to alpine. Common in southern southeastern Alaska, scattered in northern southeastern Alaska. Inner bark scraped, dried and ground into flour in spring; used locally for firewood and Christmas trees. Lives 200-600 years. Divided into two varieties:

Pinus contorta var. ***contorta*** (Shore Pine) Mostly less than 5 m tall, bark dark brown to blackish, becoming furrowed with age. Generally low spreading but can be well developed as well. Cones pointing backward and opening when mature without the influence of fire. Coastal bogs, forests and on benches near lakes. This species is shade intolerant.

Pinus contorta var. ***latifolia*** (Lodgepole Pine) Mostly greater than 6 m tall, slender, with longer needles, thin brownish bark not furrowed with age. Cones pointing outward, mostly closed for many years. Found near Skagway, Haines and Yakutat. Seeds germinate most frequently following fire since the serotinous cones readily open after being exposed to high temperatures. But in some of its range, cones open at maturity.

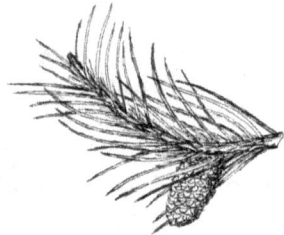

Tsuga heterophylla   Western Hemlock

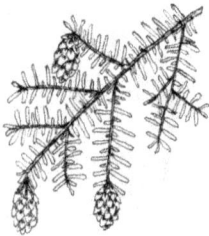

Large evergreen tree becoming 30-50 m tall and 60-200 cm in diameter with large sometimes fluted slender trunk, short narrow crown of horizontal or drooping branches, lead twig curved down or nodding, twigs reddish brown, bark reddish to gray brown becoming thick and furrowed into scaly plates, red inner bark distinguishes it from spruce. Needles spreading in two rows, rounded at tip, shiny dark green above with two rows of stomata on lower surface. Cones stalkless, hanging down at end of twig, small, elliptic with many thin papery scales. Seeds about 13 mm long including large wing.

Coastal forests from southeastern Alaska to Prince William Sound. Mostly above tidewater to subalpine. Abundant. Most common tree in southeastern Alaska. Inner bark (cambium) cut into strips and dried into cakes, new leaves edible and act like a breath mint, needles also used for tea and high in vitamin C; bark makes a red-brown dye and used as a laxative; pitch prevents chapping. Hemlock is not poisonous and did not kill Socrates. (See poison water hemlock, pg. 31) Hybrids reported between *T. heterophylla* and *T. mertensiana.* This genus has tops that curve downward and make the tree appear to be gracefully nodding. Some may live over 1,000 years.

Tsuga mertensiana   Mountain Hemlock

Small to large evergreen tree (shrub at high elevation or in peat bogs) becoming 15-30 m tall and 25-80 cm in diameter with marked taper when open grown, narrow crown of horizontal or drooping branches, leading twig nodding, bark gray to dark brown, thick and deeply furrowed. Needles mostly crowded on all sides of twigs and curved upward, flattened above and rounded or angled beneath in star-like appearance on branch, but lower branches in closed forests may appear 2-ranked as in *T. heterophylla*. Cones stalkless and usually hanging down and fives times the size of *T. heterophylla* cones, purplish but turning brown with many thin papery scales. Seeds light brown.

Southeastern and south-central Alaska in boggy areas, coastal forests and mountains. Used for railroad ties and marketed along with western hemlock as lumber. Generally slower growing than *T. heterophylla* with blockier bark. Lives 400 to 500 years.

TAXACEAE
YEW FAMILY

Taxus brevifolia   Western Yew

Small tree or large shrub to 9 m tall with straight, conical ridged and fluted trunk, bark purplish brown thin and scaly. Needles in comb-like arrangement in two rows, flat, slightly curved. Seeds single, brown, surrounded by a thick, scarlet, juicy, cup-like disk or berry.

Dog Island and surrounding islands in extreme southern southeastern Alaska in poorly drained forests and along shorelines. Rare. Berries ripen in September. The chemotherapy drug paclitaxel (taxol), used in breast, ovarian and lung cancer treatment. Western Yew is becoming increasingly scarce just as its chemotherapeutic potential is being realized. Berries reputed to be toxic: when too many eaten by Haida women, they became sterile. Heavy, close-grained wood used to make bows, wedges, clubs, paddles, fish hooks, knives, dishes, spoons, combs, snowshoe frames, etc.

27

FLOWERING PLANTS: DICOTYLEDONEAE

ACERACEAE
MAPLE FAMILY

Acer glabrum

Douglas Maple

Shrub or small tree to 10 m tall with smooth gray bark and reddish twigs, paired U-shaped leaf scars apparent during winter. Leaves opposite, deciduous, slightly heart-shaped at base, shallowly three-lobed with lobes long-pointed and sharply toothed, shiny dark-green above, pale with yellowish veins beneath, petioles slender, reddish tinged. Flowers dioecious, composed of 4 narrow yellow green sepals as long as the narrow petals, borne on 2-leaved shoots, petals yellow green. Fruit paired, winged, 1-seeded samaras not widely spreading, usually red until shed, then turning light brown.

Banks of streams, beach forests and roadsides. Occasional throughout southeastern Alaska; common in Haines area. Flowers May; fruits June-August. Green wood molded by soaking it in water and bending it while hot to form snowshoe frames. Native people tied a long branch of this maple around a red cedar and ignited it. Eventually it would burn through the trunk felling the tree. Eriophyid mites infest the leaves causing a bright red infestation known as "hummingbird's menstrual pad". Formerly known as *Acer douglasii.* This species was named to honor David Douglas, Scottish botanist and explorer. Over eighty species of plants and animals have Douglas or *douglasii* in their common and/or scientific names, in his honor, including Douglas-fir. Douglas maple could be confused with Vine Maple (*Acer circinatum*) but Vine Maple does not occur in Alaska.

APIACEAE
CARROT FAMILY

KEY TO GENERA OF *APIACEAE:*

1a. Leaves dissected into small and narrow segments, distinct leaflets not apparent; leaves ovate to triangular in outline, segments sharply pointed; stem leaves well developed; fruit 4-8 mm long.
Conioselinum

1b. Leaves not dissected into small and narrow segments, at least some with distinct leaflets. **(2)**

2a. Leaves trifoliolate, terminal leaflet mostly 10-30 cm long and broad at maturity; marginal flowers of umbel enlarged, outer corolla lobes often 2-cleft. *Heracleum*

2b. Leaves with more than 3 leaflets, usually much less than 5 cm broad; marginal flowers not enlarged, corolla lobes not 2-cleft. **(3)**

3a. Plants stemless, or nearly so; primary bracts subtending flowers lacking, secondary bracts subtending flower present; flowers stalkless; leaves densely woolly beneath; seashores. *Glehnia*

3b. Plants with well-developed, leafy stems; leaves ternately or pinnately once to several times compound; plants of various distribution. **(4)**

4a. Leaves ternately once, twice, or more times compound; leaflets mostly ovate. **(5)**

4b. Leaves pinnately once or twice compound; leaflets ovate to lance-shaped or linear.**(7)**

    5a. Leaflets mostly 9, hairless or leathery; plants maritime.
    Ligusticum

    5b. Leaflets more than 9, or if 9 or fewer, not both hairless and leathery; plants maritime or not. **(6)**

    6a. Stems slender, mostly less than 6 mm in diameter; leaves mostly once or twice compound; ovary and fruit armed with bristles or prickles. *Osmorhiza*

    6b. Stems thick, mostly over 10 mm in diameter; leaves three times compound, ovary and fruit unarmed.
    Angelica

7a. Plants decumbent to ascending, often rooting at nodes; leaflets ovate to lance-shaped; styles 1-3 mm long, persistent on fruit.
Oenanthe

7b. Plants erect, not rooting at nodes; leaflets lance-shaped to linear; styles less than 1 mm long, not conspicuous in fruit; stem base thickened, hollow, with well-developed partitions; leaves mostly more than once compound. *Cicuta*

Angelica genuflexa   **Bentleaf Angelica**

Stout perennial herb 1-1.5 m tall, transverse partitions at apex of root crown, hollow, hairless, often purple, from tuberous, woody base. Leaves divided in threes, pinnate to bipinnate, upper leaves (beyond basal pair of leaflets) abruptly bent and reflexed, leaflets broadly ovate to obovate, with sharp teeth, hairy on veins and in margins, upper sheaths of leaves inflated, veins tending to end at points of teeth. Flowers in compound umbels, inflorescence finely hairy with numerous rays. Petals small, white or pinkish. Fruit rounded, lateral ribs broadly winged, dorsal, narrowly winged.

Southeastern Alaska west to Alaska Peninsula, tip of Aleutians. Beaches, wet areas and meadows. Common. Stems used as breathing tubes under water. This species closely resembles Poison Water Hemlock since both have tuberous-thickened and chambered stem bases, but Bentleaf Angelica has kinks in its leaves and roots are not as aromatic.

Angelica lucida   **Seacoast Angelica**

Stout perennial from strong taproot, 50-140 cm tall. Leaves hairless, in threes, leaflets ovoid to triangular, acute, irregularly toothed, upper sheaths of leaves inflated. Umbel with 20-40 rays, petals white-greenish, numerous in 20 to 40 small compact heads. Ribs of fruit all similar, narrowly winged.

Southeastern Alaska west to Aleutians, central to western Alaska. Meadows, thickets, beaches, and wet areas. Common. Stem and petiole edible and called wild parsley; leaves cooked as green vegetable. Inuit inhaled fumes of roasted root for seasickness.

Cicuta douglasii Poison Water Hemlock

Stout perennial 0.5-2 m tall with stout, hollow stem, single or few together from tuberous-thickened and chambered roots. Leaves 1-3 irregularly pinnate leaflets, lance-like to narrowly oblong or elliptic, saw-toothed with forward pointing teeth, lateral veins end at base of marginal teeth rather than at the points. Flowers in several flat-topped umbels, flower bracts mostly lacking, petals white to greenish. Fruit round with low corky ribs, ribs broader than intervals.

Southeastern Alaska west to Kodiak Island and extreme eastern Alaska Peninsula, central to east-central Alaska. Marshes, beaches, along streams in and bogs. Flowers late summer. Common. All parts of the plant but especially the roots and stem base contain cicutoxin, a yellowish liquid that is **DEADLY POISONOUS** and responsible for the death of Socrates. Presence of chambered roots (Bentleaf Angelica has chambered roots as well) and primary leaf veins directed to sinuses rather than teeth of leaflets and orange-yellow resin best characteristics to distinguish Poison Hemlock. It is the most poisonous plant in North America.

Conioselinum pacificum Hemlock-parsley

Robust hairless perennial 20-120 cm tall, stem from stout taproot. Leaves all on stem, finely divided with 2-3 pinnate more or less lobed leaflets, petioles with sheath at base. Umbel with numerous rays, flower bracts few or lacking, petals white. Fruit hairless, wings of dorsal ribs narrower than lateral wings.

Mostly near coastline but sometimes in boggy forests. Common. Carrot-like taproots dug in spring, cooked and eaten. The leaves have been used to make a soothing tonic drink in the treatment of colds and sore throats and in steam baths to treat rheumatism and general weakness. Also known as *C. chinense.*

Glehnia littoralis   Beach-carrot

Coarse, fleshy, more or less prostrate maritime herb mostly less than 1 m tall, stemless or with elongate internodes buried in sand. Leaves broad, often leathery, once or twice divided into threes with rounded teeth in margins, covered in dense white woolly hairs on underside giving it silvery appearance. Compact, compound umbels in heads on stout peduncles, inflorescence woolly, petals white. Fruit rounded, flattened dorsally, mostly hairy but sometimes nearly hairless especially in margins. Southeastern Alaska west to Kodiak Island. Sandy beaches along coast. Occasional. Also known as *G. leiocarpa.*

Heracleum lanatum   Cow-parsnip

Thick, stout herb 1-3 m tall, stems hollow from thick taproot with strong odor when mature. Leaves large, broad, with dense matted hairs when young, becoming hairless, ternately compound, upper with inflated sheaths, leaflets palmately lobed, coarsely toothed, sheaths often woolly with long soft hairs at stem juncture. Umbels with 15-30 rays, terminal umbel much larger than lower, petals white. Fruit egg-to heart-shaped, lateral ribs with thin wings, dorsal ribs thread-like.

Southeastern Alaska west to Aleutians and west-central Alaska. Open areas, low elevation and alpine meadows, beaches, and along streams. Common. Marrow eaten raw and roots boiled: contains sugar and tastes like licorice; contains furanocoumarins which may cause rash in the presence of UV light in some people. Fruit and green parts pulverized to make tonics for colds and sore throat. Also known as *H. maximum*.

Ligusticum scoticum   Beach Lovage

Herb 10-80 cm tall from thick root, hairless, stem single, reddish-violet at base, lack transverse partitions within root crown. Leaves thick, long-stalked with 3 shiny and roughly toothed leaflets. Flowers in flat-topped, umbrella-like clusters, petals white or pinkish. Fruit with three ribs on back and broad intervals.

Southeastern Alaska west to Aleutians and western Alaska. Seashores. Common. Young stems and leaves eaten raw or boiled. *Ligusticum calderi* found on Grace Mt. at Dixon Entrance and on Kodiak Island.

Oenanthe sarmentosa   Water Parsley

Perennial to 1 m tall from fibrous roots, fleshy hairless stem, often reclining, branches rooting at nodes, younger shoots curled, often tendril-like. Leaves twice or imperfectly three times pinnate, leaflets cleft to toothed, hairless. Umbels usually enclosed by few narrow leafy bracts, calyx teeth lance-like, petals white. Fruit broad, with slightly raised ribs, hairless.

Southeastern Alaska west to Prince William Sound. Marshes, sluggish and brackish water. Occasional. *Oenanthe* from the Greek means wine and flower.

Osmorhiza chilensis   Western Sweetcicely

Perennial 30-100 cm tall from cluster or bundle of taproots, each leaf branch usually with 3 deeply cut leaflets. Leaflets ovate-lance-like to rounded, coarsely toothed, hairy especially along veins. Flowers erect, leafless, flower stem ends in loose umbel of tiny flowers, petals greenish-white or white. Fruit black, needle-like, 12-22 mm long, narrowing below tip, broadening to a beak, seeds sharp pointed, covered by upward pointing barbs with slender barbed tail often catching on clothing and fur.

Southeastern Alaska west to Aleutians. Forest, forest edge, shoreline forests and along seeps and streams. Occasional. Flowers May-June. Also known as *O. berteroi*.

Osmorhiza purpurea   Purple Sweetcicely

Slender perennial 20-70 cm tall from stout branched, aromatic taproot, smooth or with sparse white hairs. Leaves usually hairless, 1-3, ternate ovate-lance-shaped, incised and coarsely-toothed lobes. Flowers pink to purplish or greenish white, in umbrella-like clusters, secondary flower bracts lacking. Fruit linear-narrowed both ways from swollen middle, 8-13 mm long, constricted below shortly beaked apex with basal tail-like appendage.

Southeastern Alaska west to Kenai Peninsula and Kodiak Island. Coniferous forests, meadows and streams. Common. Flowers June-July. Fruit has an anise flavor. May hybridize with *O. chilensis*.

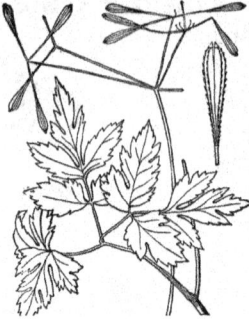

Osmorhiza depauperata (Blunt-fruit Sweetcicely) 15-70 cm tall with hairy leaves, almost apical constriction on fruit evident well before fruit maturity and observable with lens before petals fall, distinguishes this species from others; fruit club-shaped or thickened toward one end. Southeastern and south-central Alaska. Woods. Also known as *Washingtonia obtusa.*

ARALIACEAE
GINSENG FAMILY

Oplopanax horridus   Devil's Club

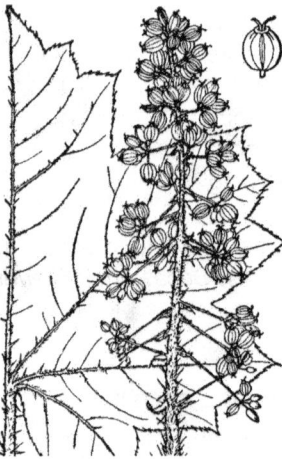

Large deciduous spiny shrub 0.5-3 m tall with several thick stems and few branches with large white pith; stems densely covered with sharp yellowish spines. Leaves few, large, alternate blades palmately veined and sharp pointed, irregularly, sharply toothed lobes, spines cover petioles and veins on lower side of leaf. Flowers numerous in terminal raceme, erect, cone-shaped, nearly stalkless, minute sepals, petals greenish white. Fruit numerous bright red berries, 1-2 seeds per berry.

Southeastern Alaska west to eastern part of Alaska Peninsula and Kodiak Island along seepages and streams, wet bottomland forest, base of cliffs, forest edge, under alder and salmonberry thickets, also in subalpine (mountain hemlock) zone. Common. Fruit may persist over winter. Berries not considered edible but commonly eaten by bears; young shoots and leaves browsed by deer. Most of plant used by native people: bark brewed for tea and used as tonic; stalk nailed over door or window to ward off evil influences; Devil's Club charcoal used for face paint; stems peeled and cut into fishing lures. Spines fester and very painful when embedded in skin. May contain an insulin-like substance and good in oil to rub on arthritic joints. Old name *Echinopanax horridum*.

ASTERACEAE
SUNFLOWER FAMILY

KEYS TO GENERA OF *ASTERACEAE:*

KEY I. COROLLAS RAY-LIKE; PLANTS WITH MILKY JUICE

1a. Plant with leafless peduncle arising from ground (scapose); heads solitary. **(2)**
1b. Plants with at least some stem leaves; heads few to several (rarely solitary). **(4)**

    2a. Achenes minutely spiny or with ridges near top of body, or less commonly smooth, tipped by slender beak.   *Taraxacum*
    2b. Achenes smooth or nearly so, beaked or beakless. **(3)**

        3a. Pappus bristles brownish, or more or less united and falling as a unit; achenes beakless.   *Apargidium*
        3b. Pappus bristles whitish, distinct, falling separately; achenes beaked.   *Agoseris*

4a. Main stem leaves hastate-deltoid, sharply toothed, petioles broadly winged; corollas white.   *Prenanthes*
4b. Main stem leaves various in shape, not as above, or much reduced upwards; corollas yellow or blue, rarely white. **(5)**

    5a. Achenes more or less flattened and beaked; stems leafy, mostly over 30 cm tall; flowers yellow or blue; cluster of bracts subtending flower cylindrical to ovoid-cylindrical.   *Lactuca*
    5b. Achenes almost circular in transverse section; stems with leaves much reduced upwards (except in some *Hieracium*), often less than 30 cm tall. **(6)**

6a. Plants from taproots; pappus white or whitish.   *Crepis*
6b. Plants from short, rhizomatous woody base; pappus tan to brown.   *Hieracium*

KEY II. RAY FLOWERS PRESENT

1a. Rays white to pink, purple, blue, or red, but not yellow. **(2)**
1b. Rays yellow or orange. **(6)**

    2a. Pappus of capillary bristles; receptacle naked. **(3)**
    2b. Pappus lacking, or a mere crown, or of scales or awns; receptacle naked or chaffy. **(5)**

        3a. Basal leaves long-petiolate, heart-shaped, arrow-shaped, or palmately lobed; stem leaves much reduced, bladeless.   *Petasites*
        3b. Basal leaves various, but not as above; stems leaves reduced or not. **(4)**
        4a. Cluster of bracts subtending flower in 3-more rows, outer ones progressively shorter, or almost equal, green throughout or more or less translucent near base; rays comparatively broad; style branches more than 0.5 mm long; plants often rhizomatous.   *Aster*
        4b. Cluster of bracts subtending flower in 1-2 rows, almost equal, green or more or less translucent throughout; rays mostly very narrow; style branches 0.5 mm long or less; plants seldom with rhizomes (except in *E. peregrinus*).   *Erigeron*

    5a. Rays short, 2-4 mm long.   *Achillea*
    5b. Rays mostly over 10 mm long; leaves entire to dentate or pinnatifid, broad lobes.   *Chrysanthemum*

FLOWERING PLANTS: DICOTS

6a. Leaves opposite. *Arnica*
6b. Leaves alternate. **(7)**
7a. Pappus of capillary bristles. **(8)**
7b. Pappus of awns, scales, a short crown, or lacking. *Chrysanthemum*

8a. Cluster of bracts subtending flower in 1 series, sometimes with 1-2 rows of very short bractlets at base. *Senecio*
8b. Cluster of bracts subtending flower in 2-more series, unequal; plants from rhizome or woody base; roots numerous; heads many, small, mostly 3-8 mm high. *Solidago*

KEY III. RAY FLOWERS LACKING OR VESTIGIAL; HEADS WITH DISK FLOWERS ONLY, ALL FLOWERS TUBULAR

1a. Pappus lacking, or of scales, awns, or a short crown. *Artemisia*
1b. Pappus of capillary or feather-like bristles. **(2)**
2a. Leaves spiny, thistle-like; receptacle densely bristly. *Circium*
2b. Leaves not spiny; receptacle not bristly (except in *Saussurea*). **(3)**
3a. Basal leaves long-petiolate, heart-shaped, arrowhead-shaped, or palmately lobed; stem leaves much reduced, bladeless. *Petasites*
3b. Basal leaves various, but not as above; stem leaves reduced or not. **(4)**
4a. Flowers pinkish or blue to purple; pappus bristles feather-like; receptacle usually bristly. *Saussurea*
4b. Flowers mostly yellow or white; pappus bristles rarely feather-like; receptacle naked. **(5)**
5a. Involucral bracts in 1 series, sometimes with 1-2 rows of very short bractlets at base
. *Senecio*
5b. Involucral bracts in 2-more series, unequal. **(6)**
6a. Basal leaves conspicuous; stem leaves much reduced; mostly pistillate plants known. *Antennaria*
6b. Basal leaves early deciduous; stem leaves little reduced, numerous; pistillate plants commonly with a few staminate flowers. *Anaphalis*

Achillea millefolium Common Yarrow

Strongly sage-scented perennial 10-100 cm tall, stem covered with white cottony hairs. Leaves fern-like, grayish-green, linear with numerous, very short, finely divided leaflets. Flowers in flat-topped clusters atop branched stems, petals white turning to pink or pink-purple with age, ray flowers usually 5, disc flowers 10-30, cream colored. Fruit compressed, hairless, achenes 1-2 mm long; pappus lacking.

Throughout Alaska except northern coast. Beaches, meadows and roadsides. Common. Younger leaves cooked as spinach or in soup. Yarrow is sweet with a slight bitter taste. Dried leaves used for tea and mixed with mint, or as an herb; also boiled and put hot on swollen infections, or good for relieving sore eyes and rheumatism. Inhaling steam from plant relieves stuffed sinuses; dried leaves used as powder for sores, cuts, or blisters, or as mosquito repellent. Also known as *A. borealis.*

Agoseris aurantiaca   Orange Agoseris

Perennial herb with milky sap 10-60 cm tall from taproot. Leaves basal, somewhat hairy, lance-shaped, entire or with widely spaced teeth. Flowers in head on leafless stem, outer cluster of bracts subtending flower head with long, dense hairs and fringed margins, often purple spotted, petals brownish orange often drying purplish, rarely yellow, consisting only of ray flowers. Achenes abruptly tapering to slender beak, half the length of the body or longer, pappus hair white.

Glacier Bay and head of Lynn Canal. Mountain meadows. Uncommon. Leaves and latex can be chewed.

Agoseris glauca   Pale Agoseris

Perennial herb 10-35 cm tall. Leaves narrowly oblanceolate, entire or less commonly toothed or lobed, sparsely hairy. Yellow ray flowers often dying pinkish with achenes lacking beaks or with a short, stout beak.

Mountain meadows and grassy slopes in Hyder area and near Skagway in southeastern Alaska. Uncommon to rare. Extremely variable species.

Anaphalis margaritacea   Pearly Everlasting

Perennial herb from 20-100 cm tall from long, creeping rhizome, stems with long, matted white hairs, leaves gradually reduced upward, alternate, broadly linear to oblong, stalkless, blunt, entire, flat to slightly curved margins; numerous oval heads with broad, pearly white papery bracts around a few disk flowers, petals off-white, achenes covered with small nipple-like bumps.

Southeastern Alaska west to Aleutians. Dry open areas, waste places and roadsides. Common. Late blooming plant whose flowers often last into winter. elated to *Antennaria*.

Antennaria umbrinella   Umber Pussytoes

Mat forming perennial herb 5-20 cm tall. Leaves basal, wedge-shaped or rounded, gray hairy, indistinctly 1-veined. Flowers in heads 3-6 almost capitate to open cymes, petals whitish or tawny, blunt involucral bracts with brownish tips. Achenes smooth.

Most of Alaska except extreme northern and western portions. Open forest mid to sub- or sometimes alpine. Occasional. Flowers late July-August. Also known as *A. isolepis* and *A. pallida*.

Antennaria alpina (Alpine Pussytoes) Closely related to *A. umbrinella* but more densely tufted and somewhat smaller, 5-10 cm tall with leafy stolons. White-wholly on both side of leaves. Flowers greenish-black with ray flowers only. Achenes mostly smooth; pappus hairs white, fine. Well drained exposed sites, gravelly ridges. Subalpine to alpine. Also known as *A. friesiana;* includes *A. media*.

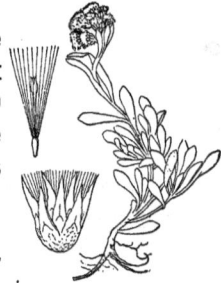

Antennaria monocephala (One-headed Pussytoes) Variable species closely related to *A. alpina*. Leaves lightly hairy above, white-woolly below. Distinguishing feature is the solitary flower head. Arctic and alpine tundra. Glacier Bay and Lynn Canal.

Antennaria pulcherrima (Showy Pussytoes) Longer and broader oblanceolate leaves and stems to 60 cm tall with 4-20 flower heads. Open woods, floodplains, and mountain slopes. Glacier Bay and head of Lynn Canal. Flowers in July.

Apargidium boreale   Apargidium

Perennial herb with milky juice 10-50 cm tall, stem from stout woody base with several taproots. Leaves basal, narrowly lance-shaped, entire or with widely spaced teeth. Flower heads solitary on scape with cluster of lance-shaped, sharp-pointed bracts subtending flower, sparsely dark-hairy, ray petals only, yellow often drying whitish or pinkish. Fruit beakless achenes 5-6 mm long, pappus has fine brownish bristles connected at base.

Southeastern Alaska west to Aleutians. Wet meadows, wet rocky slopes and sphagnum bogs. Occasional. Also known as *Microseris borealis.*

KEY TO SPECIES OF *ARNICA:*

1a. Anthers purple; bracts subtending flower callous tipped; heads nodding; leaves mostly basal.   *A. lessingii*
1b. Anthers yellow; bracts subtending flowers not callous tipped. **(2)**

2a. Stem leaves (4)5-9 pair; pappus brownish; heads 5-more per main stem (see also *A. diversifolia*). **(3)**
2b. Stem leaves (0)1-4(5) pair; pappus brownish or white; heads 1-4(5) per main stem. **(4)**

     3a. Cluster of bracts subtending flower acuminate to attenuate, lacking apical tuft of hair.   *A. amplexicaulis*
     3b. Cluster of bracts subtending flower merely acute to abruptly rounded (rarely acuminate), bearing apical or almost apical tuft of hair.   *A. chamissonis*

4a. Leaves (at least the lower) heart-shaped, ovate, of broadly lance-shaped, heart-shaped, truncate or obtuse basally, seldom cuneate. **(5)**
4b. Leaves narrowly lance-shaped to lance-oblong, usually cuneate basally. *A. mollis*

     5a. Pappus brownish, more or less feather-like; leaves usually obtuse to almost wedge-shaped basally.   *A. diversifolia*
     5b. Pappus white, merely finely barbed; leaves usually almost heart-shaped, truncate, or obtuse basally (at least lower). **(6)**

6a. Achenes hairless, at least near base, blades of basal leaves much longer than petioles, or petioles lacking; plants of southern and southeastern Alaska.   *A. latifolia*
6b. Achenes uniformly though sometimes sparingly hairy; blades of basal leaves almost equal to, or shorter than petioles; plants of southeastern Alaska.   *A. cordifolia*

Arnica amplexicaulis   Clasping Arnica

Perennial herb to 80 cm tall from coarse creeping rhizome, glandular and hairy especially upward. Stem leaves in several pairs, stalkless, lance- to egg-shaped, irregularly toothed. Flowers in several heads, involucral bracts lance-shaped, sharp pointed, hairy and sparsely glandular at base, few short glands at tip, ray flowers pale yellow, disc flowers yellow. Achenes sparsely hairy, brown, feathery pappus.

Southeastern Alaska west to Aleutians. Along streams, wet areas, open forests, mountain slopes, subalpine meadows and lake shores. Low to mid elevation. Occasional.

Arnica chamissonis   Leafy Arnica

Perennial herb 20-100 cm tall, stems solitary from long slender rhizome. Leaves mostly stalkless, lance-shaped or ovate-lance-shaped, slightly toothed or entire margin, flower bracts hairy. Flower ligules pale yellow, disk flowers covered with coarse shaggy hairs, petals yellow. Achenes hairless or sparsely hairy.

Southeastern Alaska west to Aleutians. Meadows, open areas. Common.

Arnica cordifolia   Heart-leaved Arnica

Perennial herb 20-60 cm tall, stems solitary or few loosely clustered together, glandular with short soft hairs to loosely white-hairy. Leaves heart-shaped or orbicular becoming lance-shaped above, on long petioles, stem leaves mostly in 2-4 pairs, lower pair larger. Flowers 1-3 or sometimes more heads with toothed rays, involucrum sparsely to densely covered with long spreading white hairs, often glandular as well, white pappus, petals yellow. Achenes with short hairs or glandular.

Southeastern Alaska. Open to dense forests. Occasional. Flowers May-June. More hairy involucral bracts and achenes and more heart-shaped leaves distinguish this species from **A. latifolia** which is closely related.

Arnica latifolia   Mountain Arnica

Perennial herb 10-60 cm tall, stem mostly solitary from scaly horizontal rhizome, appears more leafy than *A. cordifolia*. Basal leaves ovate to lance-shaped, rarely heart-shaped, smaller than stem leaves, soon withering; stem leaves in 2-4 pairs, stalkless with more teeth than basal leaves and similar in shape. Flowers in 1-several heads, smaller than those of *A. cordifolia*, cluster of bracts subtending flower with stalked glands, petals yellow. Achenes smooth, pappus white, finely barbed.

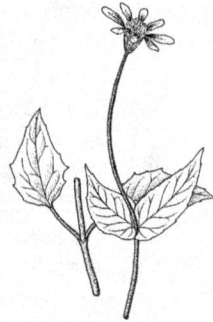

Southeastern Alaska west to Aleutians. Open areas and meadows. Common. *Arnica* is commonly used medicinally in Europe as an antiseptic and if taken internally elevates body temperature.

Arnica lessingii   Slender Arnica

Perennial herb simple or rarely branched above, 10-50 cm tall, moderately to densely covered with multicellular hairs, basal leaves smaller than stem leaves, often withered at flowering time, or all leaves appearing basal, broadly winged petiole; stem leaves in 3-5 pairs, lance-shaped to elliptic, entire to sharply toothed. Flower heads solitary and nodding with conspicuous teeth and yellow rays, purplish anthers. Achenes sparsely hairy to hairless, pappus brownish.

Most of Alaska. Alpine meadows. Occasional. Flowers late July-August. May be hybridization of *A. cordifolia* and *A. latifolia*.

Arnica diversifolia (Diverse Arnica) Shallowly toothed yellow ray flowers and glandular hairs on leaves; pappus brownish. Open woods of southern Alaska. May represent a complex hybridization among *A. amplexicaulis, A. mollis, A. cordifolia* and *A. latifolia*. Uncommon.

Arnica mollis (White-haired Arnica) Perennial herb 15-60 cm tall, stems solitary with broader leaves, sparsely to moderately hairy, basal leaves elliptic, smaller than stem leaves. Flower heads 1-5, peduncle apex sparsely white-hairy; rays yellow. Alpine meadows and mountain slopes. Reported from Stikine River and Shakes Lake near Petersburg.

41

Artemisia norvegica   Boreal Wormwood

Perennial herb 20-60 cm tall, stems from stout, woody base with short runners, sterile rosettes and persistent leaf bases, stems sometimes reddish, flower stems arise directly from woody base. Basal leaves bipinnately dissected into 5-7 pairs of lobes, ultimate lobes linear to linear lance-shaped, acute; stem leaves reduced, often with stipule-like divisions near base, upper leaves linear. Flower heads nodding, lower ones on long peduncles, bracts hairless or nearly so, greenish, with thin, dry, membranous brownish margins, corolla long-hairy from near base, petals yellowish, sometimes tinged red. Achenes hairless, pappus lacking.

Most of Alaska. Rocky places in alpine meadows, gravelly soils and rocky ridges. Occasional to common. This species is food for a number of animals, such as mountain goats, Sitka black-tailed deer and hoary marmots. Also known as *A. arctica*.

Artemisia tilesii   Aleutian Mugwort

Perennial herb, stems often purplish from stout rhizome and fibrous roots, 30-150 cm tall. Leaves primarily along stem, hairy below, usually green and shiny above, once or twice pinnatifid, entire to more deeply lobed with forward-pointing lobes. Flower heads nodding to spreading, petals yellow often tinged reddish. Achenes hairless.

Most of Alaska. Rocky or gravelly areas mostly high in mountains but also along roads. Raw shoots peeled and eaten with oil; fresh leaves used as a flavoring, also made into tea for colds; leaves and flowering tops infused and used as a laxative and to treat stomach aches, rheumatism, chest colds, or applied externally to swollen joints. A poultice of the leaves is also used to treat skin infection and cuts, stop infection and promote healing. Flowering heads can be used to make healing and soothing ointment and in steam bath for the fragrance and medical qualities; fresh leaves used to clean odors from hands and in houses to refresh the air and as mosquito repellent.

Artemisia campestris (Northern Wormwood) Leaves green, usually with densely silky hairs on both surfaces, ultimate segments becoming linear; flower heads erect, disk corolla yellow and sparsely hairy to hairless. In our area occurs only on mainland of southeastern Alaska from Petersburg to Lynn Canal. Alpine tundra. Highly variable.

42

Aster modestus   Great Northern Aster

Perennial herb 40-100 cm tall, stems single from creeping rhizome, glandular above. Basal leaves usually withering by flowering time; stem leaves lance-shaped, coming to sharp point with widely spaced teeth, upper leaves more or less clasping. Bracts subtending flowers narrowly lance-shaped and sharply pointed, heads several to numerous, glandular, rays violet or purple. Achenes several nerved, sparsely hairy, pappus hairs whitish or yellowish.

Southeastern Alaska. Open forests and along streams. Uncommon. Asters often flower very late in the season. Also known as *Canadanthus modestus.*

Aster subspicatus   Douglas Aster

Perennial herb 20-80 cm tall, stem from creeping rhizome, plant sparsely hairy. Basal leaves smaller than stem leaves, usually withered by flowering time; stem leaves stalkless, slightly clasping, lance-shaped to oblong, to narrowly elliptic, hairless above, with short, rough hairs below. Flower heads solitary or more commonly few to several in clusters, peduncles with long, soft, wavy hairs, bracts oblong-lanceolate, hairless beneath, short hairs on margins, partly overlapping in several series, green at tip, margins near base brownish or yellowish, rays purple, pappus brownish or purplish, rarely whitish. Achenes hairy nerved, pappus hairs reddish or purplish-brown.

Southern Alaska. Open areas, meadows and roadsides. Common. Extremely variable, forms hybrids with species to south. Also known as *A. douglasii* and *Symphyotrichum subspicatum.*

Chrysanthemum arcticum   Arctic Daisy

Perennial herb 20-80 cm tall, stem from thick, creeping woody base with the odor of sage. Basal leaves spoon-shaped, pinnatifid, cobwebby hairs at base with 3-7 blunt often toothed lobes; stem leaves reduced, upper ones linear becoming entire. Flower heads usually solitary, bracts subtending flowers green with brown translucent margins, rays of marginal flowers several-nerved, rays white, disk flowers yellow. Achenes almost circular in transverse section, about 10-ribbed, pappus lacking.

Coastal Alaska. Beaches, meadows and on rocks along seashore. Common. Flowers June-September. Also known as *Dendranthema arcticum* and *Leucanthemum vulgare*.

Cirsium edule   Edible Thistle

Stout, spiny herb from taproot, stems mostly 50-200 cm tall with cobwebby hairs. Leaves shallowly or deeply pinnatifid, hairless except on midrib below, spiny. Flowers solitary or few, bracts lance-linear at least outer ones spine tipped, petals pink purple, achenes 4-angled, hairless, pappus with feathery, white bristles.

Reported in extreme southeastern Alaska from Hyder. Streamsides and wet meadows. Uncommon. Flowers July-August. Roasted roots have sweet taste. Roots and leaves used as talisman.

Crepis nana   Dwarf Hawk'sbeard

Dwarf, densely tufted herb with milky juice and much branched stems to 10 cm tall. Leaves mainly basal, numerous, flattened spoon-shaped to orbicular or ovate, hairless. Flowers in several to many heads, borne in compact, cushion-like clusters, petals yellow. Achenes 4-6 mm long, golden brown, ribs smooth.

Northern southeastern Alaska. Alpine. Open gravelly places, rivers and lake shores. Occasional. Stems and petioles elongated when covered with detritus. Another species *C. elegans* is found on gravelly or sandy river bars in northern southeastern Alaska.

Erigeron humilis   Arctic-alpine Daisy

Perennial herb 3-20 cm tall, upper part of plant covered by long soft purple-black hairs. Basal leaves long-stalked in rosette; stem leaves reduced and linear. Flower heads solitary (rarely 2), petals white to pink or purplish, disk corolla almost equal to or exceeded by white to tan pappus. Achenes 2-nerved, hairy.

Northern southeastern throughout Alaska. Gravelly soil in alpine meadows, sometimes bogs at low elevation. Occasional to common.

Erigeron peregrinus   Wandering Daisy

Perennial herb 10-70 cm tall, more or less hairy single stem from short stout woody base. Leafy, margins fringed with white hairs, basal leaves narrowly or broadly lance-to flattened spoon-shaped; stem leaves lance-shaped to oblong, stalkless, smaller than basal leaves. Flower heads solitary (rarely more), bracts subtending flowers hairy, 30 to 80 strap-like flowers, disk corolla exceeding tan pappus, petals pink, purple, or white. Achenes 4-7-nerved, hairy, pappus hairs white to tan.

Southeastern Alaska west to Aleutians; less common in Interior. Wet meadows, open woods, bogs and alpine. Common. Flowers June-September.

Erigeron compositus (Cut-leaved Daisy) Perennial herb forming dense tufts with crowded once or twice ternately divided leaves with spreading hairs and often glandular as well, stem leaves greatly reduced; solitary flower heads with purple-tinged bracts, ligules white, pink, blue, or lacking, disk corollas exceed whitish to tan pappus; achenes 2-nerved and hairy. Open slopes, rock outcrops, and lake shores mid to high elevation from Haines area through eastern half of mainland Alaska, disjunct in northwestern Alaska.

Erigeron lonchophyllus (Short-rayed Daisy) Sparsely to moderately spreading-hairy, basal leaves oblanceolate tapering to petiole, stem leaves linear, stalkless, often with long hairs on margins; flower heads solitary or more commonly few to several in raceme-like clusters, rays pink or white, disk corollas shorter than white or whitish pappus; achenes 2-nerved and hairy. Dry open slopes, gravel bars and bogs in Glacier Bay. Flowers in July.

Hieracium albiflorum   White-flowered Hawkweed

Perennial herb 30-100 cm tall with milky juice, stems 1-several, with long, whitish or tawny spreading hairs near base, hairless or nearly so above, erect, tall, leafy below. Leaves sparsely to moderately covered with long hairs, oblong to broadly lance-shaped, upper leaves small. Flower heads several to many, slender peduncles sparsely long-hairy and glandular, petals white. Achenes 2-3 mm long, pappus bristles tawny.

Southeastern Alaska. Open forests, meadows and clearings. Low to mid elevation. Occasional. Flowers June-August. Some hawkweeds, such as *H. aurantiacum*, were introduced to Alaska and are serious invasive weeds in the region, but the 3 hawkweeds described on this page are native.

Hieracium gracile   Slender Alpine Hawkweed

Perennial herb 8-25 cm tall, similar to *H. triste* but more slender and often shorter. Leaves oblanceolate, entire or toothed margins, tapering basally to slender petiole, sparsely hairy, glandular, stem leaves much reduced above. Flower heads solitary or more commonly few to many, peduncles white-hairy with blackish glands on stalks, petals yellow. Achenes 2.5-3.5 mm long, pappus off white to straw-colored.

Southeastern Alaska west to eastern Aleutians. Alpine, open slopes, and disturbed areas. Common. Flowers July. Forms hybrids with *H. triste*.

Hieracium triste   Woolly Hawkweed

Perennial herb 1-40 cm tall, stems densely gray hairy. Leaves obovate to flattened spoon-shaped, entire or remotely toothed margins tapering basally to slender petiole, margins fringed with long and short hairs. Flowers solitary or more commonly 2-5(7), peduncles with long blackish or grayish hairs underlain by whitish hairs, muddy colored pappus, sometimes not fully open, petals yellow. Achenes angular, beakless, 2-3 mm long.

Southern Alaska. Rocky and sandy slopes in alpine. Occasional. Flowers July-August. Forms hybrids with *H. gracile*. Native hawkweeds can be distinguished from introduced species by their smaller flowerheads, less than 9.5 mm.

Petasites frigidus   Arctic Sweet Coltsfoot

Perennial herb 10-50 cm tall, stems simple from thick, creeping rhizome, stem scales reddish. Leaves long-stalked, triangular to heart-shaped, with 5-8 broad teeth, hairless above, white hairy below. Flowering stems precede leaves, flower heads several to many, peduncles often densely woolly with glands on slender stalk-like base, rays small to conspicuous, petals white. Achenes 5-10-nerved, hairless, beakless, pappus with numerous white hairs.

Throughout Alaska. Open areas, open forest and along streams. Subalpine to alpine. Common. Flowers June-August. Young leaves, stalks, and flower heads edible; soaked root used for sore throat, tuberculosis and stomach ulcers; applied to insect bites, swelling and phlebitis; large doses may cause abortion. Variable species which forms hybrids. Also known as *P. nivalis* or *P. hyperboreus.*

Prenanthes alata   Western Rattlesnake-root

Tall perennial herb 20-80 cm tall, hairy with milky juice and single stem. Leaves mostly on stem with winged stalks, white below, arrowhead-shaped, irregularly toothed, upper reduced. Flowers clustered, heads few to many, ligules white, aging to purple, toothed at apex with styles longer than ligules, petals white or pinkish. Achenes 5-15-nerved, beakless, hairless; pappus brown.

Southeastern Alaska west to Aleutians. Forests, beach fringe and subalpine forests, streamsides and roadsides. Common. Flowers July-September.

Saussurea americana   American Sawwort

Perennial herb 50-90 cm tall, stem tall and coarse from stout woody base. Leaves lance-shaped-ovate, coarsely toothed, lower stalks with soft woolly hairs when young. Inflorescence dense, bracts subtending flowers firm, ovate-lance-shaped with dark edges, petals pink-purple or white. Pappus white to brownish.

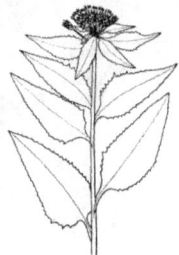

Juneau area north to Haines and Skagway, along coast near Yakutat. Moist meadows and open woods. Uncommon. Flowers July-August.

KEY TO SPECIES OF *SENECIO:*

1a. Stems, leaves, and cluster of bracts subtending flower sparsely to moderately or densely woolly, bracts (20)25-55 mm broad; stem leaves well developed; plants of coastal beaches.   *S. pseudo-arnica*

1b. Stems, leaves, and cluster of bracts subtending flower variously pubescent or hairless, bracts commonly less than 25 mm broad; habitat seldom as above. **(2)**

2a. Main stem leaves stalked, blades hastate-triangular to ovate, or lance-shaped, sharply toothed, base hastately lobed, nearly straight across, or abruptly obtuse to acute.   *S. triangularis*

2b. Main stem leaves stalkless, or if stalked, then blades lance-shaped to oblong or oblanceolate, entire, finely to sharply toothed, or pinnatifid, attenuate to acute basally. **(3)**

3a. Heads solitary (rarely 2); blades of basal leaves about as broad as long or broader, teeth merely rounded (rarely lobed); plants hairless or nearly so.   *S. resedifolius*

3b. Heads 2-many (rarely solitary); blades of basal leaves longer than broad, sharply to round toothed, or lobed; plants hairless, or more or less densely woolly. **(4)**

4a. Leaves with fine sharp teeth, main stem leaves ovate to lance-shaped, lance-oblong, elliptic, oblanceolate, or triangular, often more than 6 cm long.   *S. lugens*

4b. Leaves with round to minute teeth, or pinnatifid, main stem leaves linear to lance-oblong, oblanceolate, or elliptic, often less than 6 cm long. **(5)**

5a. Disk corollas yellow; heads commonly 8-20; leaves thin; plants in lowlands of continental Alaska.   *S. indecorus*

5b. Disk corollas orange or reddish orange; heads commonly 1-6; leaves somewhat succulent; plants alpine or subalpine; reported in Glacier Bay.   *S. pauciflorus*

Senecio pseudoarnica   **False Arnica**

Robust perennial herb 10-70 cm tall from deep taproot, stem stout with white woolly hairs. Leaves spoon- or egg-shaped to oblong, fleshy green above, white hairy below, wavy toothed. Flower heads large and few, bracts subtending flowers more or less hairy, ligules yellow, pappus white, ray and disk flowers yellow. Achenes hairless.

Southeastern Alaska west to Aleutians and west coast of Alaska. Sandy or gravelly beaches and tide flats. Common. Flowers July-August. Fleshy stems and boiled leaves used for food. Only large sunflower-like plant found along beaches in our area.

Senecio triangularis Arrow-leaved Groundsel

Perennial herb 30-120 cm tall, stems several, simple, from fibrous-rooted stem-base or rhizome. Basal leaves lacking, stem leaves gradually reduced upwards, triangular-heart-shaped, squared off at base, sharply toothed. Flower heads several in flat-topped inflorescence, bracts subtending flowers lance-shaped, glandular at tip, ligules few, ray and disk flowers yellow. Achenes hairless.

Southeastern Alaska west to Alaska Peninsula. Open areas and meadows especially in subalpine. Common. Flowers July-September. *Senecio* is one of the largest and most wide ranging genera of flowering plants in the world.

Senecio indecorus (Rayless Mountain Butterweed) 10-80 cm tall, stem hairless, basal leaves elliptic to obovate or lance-shaped, rounded to sharp toothed, hairless, with slender petioles, usually densely woolly near base, stem leaves coarsely toothed, reduced upwards, eventually stalkless with clasping ear-like appendages; flower heads 4-many in open, almost umbel-like corymbs, peduncles with 1-several small bractlets, bracts in 1 series or with a few short outer scales, lance-linear, green or tips tinged with purple, ray flowers lacking (rarely present, then yellow), disk flowers yellow; achenes hairless. Open slopes, open woods and roadsides from Yakutat throughout continental Alaska. Also known as *Packera indecora.*

Senecio lugens (Black-tipped Butterweed) 10-80 cm tall, usually from solitary stem becoming hairless, basal leaves lance-shaped to elliptic, toothed or nearly entire, few reduced stalkless stem leaves; few to many flower heads in compact to open cymes, peduncles densely woolly, with bracts in two series, outer ones much shorter, green, with black triangular tips, rays yellow; achenes hairless, pappus white. Meadows, bogs and open woods in most of our range except extreme southeastern and southwestern Alaska. Flowers June-July. *S. moresbiensis* found in southern southeastern Alaska has egg-shaped regularly toothed, purplish basal leaves and purplish involucral bracts.

49

Senecio pauciflorus (Few-flowered Groundsel) 15-30 cm tall, stem hairless, leaves mainly basal, roundish to elliptic-ovate with small round teeth, stem leaves reduced in size, toothed to pinnatifid; flower heads few, lacking ligules, flowers orange or reddish, bracts subtending flower reddish to purple, lack ray flowers; achenes hairless. Alpine and subalpine meadows and moist cliffs at head of Lynn Canal, Glacier Bay and Prince William Sound. Flowers July-August. Also known as *Packera pauciflora.*

Senecio resedifolius (Dwarf Arctic Butterweed) 1-20 cm tall, nearly hairless from simple stem, basal leaves ovate to kidney-shaped, rounded to sharp toothed, stem leaves mostly pinnatifid, upper ones reduced; flower heads solitary, bracts often purplish, rays long, yellow, mostly reddish on back or lacking, disk flowers orange; achenes hairless, pappus white. Arctic and alpine tundra in most of Alaska. Also known as *Packera cymbalaria*.

Solidago canadensis   Canadian Goldenrod

Perennial herb 40-150 cm tall, stem from long, creeping rhizome, densely leafy and thinly haired for entire length. Leaves numerous, reduced in size beneath flower, lance-shaped with rough margin, 3-nerved, outer margins sharply toothed. Flowers in broad panicle with dense thread-like racemes, bracts lance-linear, petals yellow; achenes hairy.

Southeastern Alaska west to Alaska Peninsula and Interior Alaska. Meadows, shorelines and open woods. Common. Flowers July-September. All *Solidago* species contain some natural rubber in their latex. Also known as *S. lepida*.

Solidago multiradiata Northern Goldenrod

Perennial herb 5-40 cm tall, erect hairy stem from short or rarely elongate rhizome. Basal leaves lance- to flattened spoon-shaped becoming oval near top of stem, leaf petiole fringed with hairs. Flowers in dense heads, ray flowers longer than disk flowers, petals yellow. Achenes with soft spreading hairs, pappus hairs white.

Throughout Alaska. Meadows, open areas from lowlands to lower alpine region. Common. Flowers late June-August. This is one of the most common and widespread species in the alpine tundra.

Taraxacum ceratophorum Horned Dandelion

Dwarf herb mostly to 50 cm tall, similar to common weedy dandelion but less robust and native to our area. Leaves various in form, most commonly strap-like, 4-35 cm long, 0.7-6 cm broad. Flower stems thickened, gray flower head bracts broadly rectangular, bracts, especially inner, bearing little horns or crests, rays yellow or fading to blue. Achenes 3-7 mm long, straw-colored to brownish, beak 2-3 times longer than body, pappus white.

Juneau north throughout Alaska. Meadows and bogs from mid to high elevation. Occasional. Flowers July-August. This is a complex of species which includes *T. trigonolobum*, *T. lyratum* and *T. eriophorum*.

BALSAMINACEAE
JEWELWEED FAMILY

Impatiens noli-tangere Western Touch-me-not

Succulent annual herb 20-80 cm tall, generally hairless, stem watery to fleshy. Leaves alternate, egg-shaped, wedge-shaped at base with coarse rounded teeth. Flowers irregular, showy, with 3 sepals, upper one petal-like and spurred, petals 5, appearing as 3, bright yellow spotted reddish brown or purple, spurred sepal yellow, spotted purplish, stamens 5, filaments short, anthers fused above stigma, ovary 1, superior, generally 5-chambered, stigmas 1-5, stalkless; fruit capsule, 5-chambered; explodes at touch when ripe.

Southern Alaska except Aleutians. Wet meadows, stream banks, deciduous forests, beaches and along trails. Occasional. Flowers July-September.

BETULACEAE
BIRCH FAMILY

Alnus incana   Thinleaf Alder

Deciduous large shrub or small tree to 9 m tall, trunk to 20 cm in diameter; twigs reddish and hairy when young, becoming gray, bark gray to dark gray, smooth, becoming reddish gray, thin and scaly. Leaves ovate or elliptic, short-pointed, rounded at base, shallowly wavy lobed and doubly toothed with large and small teeth, thin, dark green becoming hairless above. Male flowers in narrow catkins. Fruit cone-like on short stalks, nutlets elliptic, wingless, appearing before leaves.

North of Juneau to Glacier Bay, base of Alaska Peninsula, Kenai Peninsula and Copper River Valley, low to mid elevation. Forms thickets with larger willows along streams. Flowers May-June. Large trunks used for poles; fire made with alder wood good for smoking salmon. Also known as *A. tenuifolia*.

Alnus rubra   Red Alder

Small to medium sized deciduous tree with straight trunk, twigs hairy when young, becoming dark red with light dots, bark gray, often splotched with white crustose lichen, smooth, becoming slightly scaly and thin, winter buds stalked and dark red. Leaves ovate to elliptic, short pointed at both ends, shallowly wavy lobed with coarse, blunt teeth, thick, edges slightly curved under, rusty hairs along veins. Male flowers in narrow catkins. Fruit on short stalks, cone-like, nutlets elliptic with 2 narrow wings.

Southeastern Alaska north to Glacier and Yakutat bays. Forest edges, open areas, wetlands, along streams and roadsides, clearcuts, and following retreat of glaciers, generally at low elevation. Common from Juneau south, occasional north to Yakutat. National Cancer Institute found this species to contain two anti-cancer agents. Inner bark and wood turn rusty-red when cut. Wood used as fuel and for smoking fish since it is low in pitch and does not impart flavor to fish; reddish dye made from bark. Important nitrogen fixer improving fertility of soil. Also known as *A. crispa* or *A. oregana*.

Alnus crispa   Sitka Alder

Deciduous shrub or small tree to 9 m tall with sticky finely hairy twig, orange brown when young, becoming light gray; bark gray to light gray, smooth and thin, stems generally curved and clumped. Leaves ovate, short pointed, rounded at base, shallowly wavy lobed and doubly toothed with long-pointed teeth of two sizes, sticky when young, speckled yellow green above. Male flowers in narrow catkins. Fruit on long slender, spreading stalks, nutlets elliptic, with 2 broad wings.

Most of Alaska. Pioneer species forming thickets on mountain slopes, avalanche tracts, along stream banks and shorelines, terraces, and on recently deglaciated areas. Common to abundant. Flowers May-June. Cambium eaten or made into tea and used to reduce fever or as gargle for sore throat; female flower clusters taken in small quantities for relief of diarrhea. Curved stems allow Sitka Alder to spring back after burial by snow and mud so it persists on unstable slopes. Also known as *A. sinuata* or *A viridis.*

Betula papyrifera   Paper Birch

Small to medium-sized deciduous tree usually to 30 m tall and 10-30 cm in diameter, bark creamy white to copper-brown, coming off in thin layers. Leaves deciduous, oval to round, sharp-pointed, margins doubly toothed. Male and female flowers borne in separate catkins 2-4 cm long which fall apart when mature, emerging at the same time or before the leaves.

Open dense woods, marine shoreline, poor rocky sites and wetlands. Sap used to make syrup, wine and health tonics, tree itself for baskets and canoes, resin as chewing gum.

Two subspecies in our area: *B. p. commutata* occurs uncommonly along large streams on mainland southeastern Alaska from Petersburg to head of Lynn Canal. Bark reddish brown, trunks often clustered having originated from sprouts at base of old trees. *B. p. kenaica* occurs on Kenai Peninsula and Kodiak Island.

Betula glandulosa   Glandular Birch

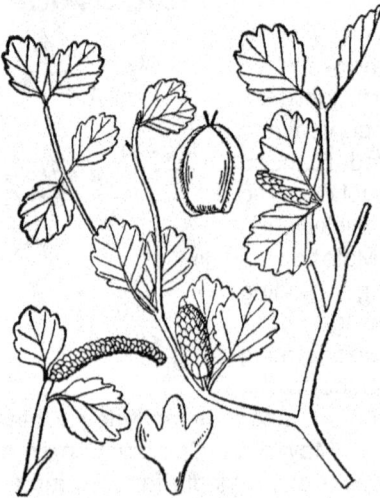

Low spreading deciduous shrub to 1.5 m tall, twigs finely hairy when young, densely resinous with warty glands and gray layer of wax. Leaves alternate, almost stalkless or with slender petioles, blades round or kidney-shaped, often broader than long, fine wavy toothed margins, thick, hairless, turning copper red in fall. Male flower clusters with brown scales, female clusters greenish. Fruit cone-like, mostly erect, with many 3-lobed bracts or scales with resinous dot or hump on back.

Northern Lynn Canal, Yakutat to Kenai Peninsula, Kodiak Island, Alaska Peninsula and most of Interior Alaska. Bogs, wet open areas, rocky alpine slopes. Common. Flowers June; fruits July and August. Includes *B. nana*. *Betula* species commonly hybridize and are often difficult to distinguish.

BORAGINACEAE
BORAGE FAMILY

Amsinckia menziesii   Small-flowered Fiddleneck

Annual from taproot, 15-80 cm tall, simple or branched with spreading hairs on blister-like bases. Leaves alternate, entire, linear to oblong, or lance-shaped, stiffly hairy and basally clasping, lower ones stalked. Flowers in coiled false-raceme, corolla regular, 5-lobed, light yellow, sepals 5, free, stiffly hairy, stamens 5, inserted above middle of tube. Fruit 4 ovoid nutlets, egg-shaped, gray to black, wrinkled and warty dorsally.

Southern Alaska where it may represent recent introduction from western Canada. Flowers unfold like the fiddlehead of a fern. Open areas and roadsides. Flowers June-September.

Lappula echinata   Stickseed

Annual or biennial coarse herb to 45 cm tall. More densely pubescent than *Amsinckia.* Flowers tiny 5-petaled blue or white with yellow center, in loose clusters near top of plant, sepals hairy. Fruits large nutlets on erect or ascending stalks with marginal hooked prickles in 2-3 rows, dorsal surface with pimple-like structures, not hairy.

Weedy species of roadsides and well-drained hillsides in southern Alaska. The hooked seeds stick to passing fur and clothing, hence the common name. Flowers July-August. Also known as *L. myosotis*.

Mertensia maritima   Oysterleaf

Herb 10-100 cm tall from branched, often subterranean woody base and prostrate taproot, hairless throughout. Leaves fleshy, alternate, ovate to obovate or elliptic, hairless on upper surface with slight, blister-like bumps. Flowers several to many in almost flat-topped cluster with outer flowers opening first, petals blue, rarely white, pink in bud, united in short basal tube, flares into 5 lobes, sepals 5, hairless. Fruit 4 clustered nutlets, smooth or slightly wrinkled.

Coasts and islands of Alaska. Sand and gravel ocean beaches. Common. Flowers June-early August. Long, leafy stems boiled, cooked briefly and eaten with seal oil. Rhizome also edible.

Plagiobothrys scouleri Scouler's Popcornflower

Annual to 20 cm long from taproot with appressed hairs, erect or prostrate; leaves along stem, linear, covered with stiff hairs, alternate, lower 1-4 pairs opposite. Flowers in loose spikes at end of branches, globular, petals white, calyx 2-4 mm long with 5 lance-shaped lobes; corolla tube narrow, the 5 lobes abruptly spreading, limb 1-4 wide, white; hairy appendages opposite the corolla lobes at the top of tube; sepals 5. Fruit wrinkled ovoid nutlets.

Weedy species of wet, disturbed sites in southeastern Alaska. Characteristic of the Borage Family, the flowers open starting from the bottom of the plant moving toward the top. This causes the stem to uncoil. Also known as *P. cognatus*.

BRASSICACEAE
MUSTARD FAMILY

KEY TO GENERA OF *BRASSICACEAE:*
1a. Plants with stem leaves that clasp stem. **(2)**
1b. Plants with stem leaves stalkless or stalked, but not clasping, or leaves basal. **(5)**
  2a. Flowers yellow. **(3)**
  2b. Flowers white, pink, or purplish. **(4)**
3a. Pods 1-3 times longer than broad (silicle); stems spreading-hairy, at least below.
Rorippa
3b. Pods more than 5 times longer than broad (silique); stems hairless or becoming hairless. *Barbarea*
  4a. Pods several times longer than broad; petals white, pinkish, or purplish; plants hairy at least below, or more than 20 cm tall. *Arabis*
  4b. Pods mostly less than twice longer than broad; petals white, plant hairless, less than 20 cm tall. *Thlaspi*
5a. Plants aquatic or maritime, fleshy, or terrestrial and petals lacking. **(6)**
5b. Plants terrestrial; seldom aquatic or maritime, rarely fleshy; petals present. **(11)**
6a. Plants strictly aquatic. **(7)**
6b. Plants terrestrial or maritime, but not aquatic. **(8)**
  7a. Leaves simple, linear, more or less circular in transverse section.
Subularia
  7b. Leaves compound or pinnatifid, blade well developed, flat. *Rorripa*
8a. Fleshy maritime herbs; basal leaves with long, slender petioles, blades kidney-shaped to oval, or wedge- to spatula-shaped with coarse, sharp, forward pointing apical teeth. **(9)**
8b. Seldom fleshy or maritime, but if so (compare *Cakile*), leaves not as above. **(10)**
9a. Basal leaves with slender petioles, blades kidney-shaped to oval; flowers white; silicles 4-7(10) mm long. *Cochlearia*
9b. Basal leaves wedge- to spatula-shaped; flowers yellow; silicles (6)8-25 mm long.
Draba

10a. Petals lacking; silicles flattened at right angles to partition, about as broad as long. *Lepidium*

10b. Petals present; silicles with 2 parts, breaking into an upper and a lower segment at maturity, longer than broad. *Cakile*

11a. Flowers yellow. **(12)**

11b. Flowers white, pink, or purple. **(15)**

12a. Fruit not over 3 times longer than broad. **(13)**

12b. Fruit over 3 times longer than broad. **(14)**

13a. Fruit distinctly compressed parallel to partition; elliptic to oblong in outline. *Draba*

13b. Fruit circular in transverse section or quadrangular, slightly if at all flattened. *Rorippa*

14a. Fruit at maturity strongly compressed parallel to partition, often twisted. *Draba*

14b. Fruit circular in transverse section or quadrangular, not strongly compressed. *Erysimum*

15a. Fruit body not over 3 times longer than broad (somewhat longer in *Cakile*). **(16)**

15b. Fruit body 3-many times longer than broad. **(18)**

16a. Plants succulent maritime annuals; fruit transversely 2-segmented. *Cakile*

16b. Plants of various habitats, usually not maritime, not succulent; fruit not 2-segmented. **(17)**

17a. Fruit flattened at right angles to partition. *Lepidium*

17b. Fruit not flattened, or flattened parallel to partition. *Draba*

18a. Plants at least somewhat hairy, with some or all hairs starlike or branched. **(19)**

18b. Plants hairless, or with simple hairs only (glandular-pubescent in *Phoenicaulis*). **(22)**

19a. Fruit at maturity strongly compressed parallel to partition. **(20)**

19b. Fruit circular in transverse section or quadrangular, not strongly compressed. *Arabis*

20a. Petals bright pink to pink purple; siliques 25 mm long or more, 2-3 mm broad. *Erysimum*

20b. Petals white or yellowish, 2-7 mm long; siliques 5-45 mm long, less than 1 mm broad, or if broader, than less than 20 mm long. **(21)**

21a. Fruit less than 20 mm long, often twisted, elliptic to oblong or lance-shaped. *Draba*

21b. Fruit over 20 mm long, not twisted, oblong. *Arabis*

22a. Fruit circular in transverse section or nearly so; plants of wet places, rooting at nodes. *Rorripa*

22b. Fruit distinctly flattened; plants of various habitat, if growing in wet places, then not rooting at nodes (except in some *Cardamine*). **(23)**

23a. Plants glandular-pubescent throughout, rarely hairless; siliques 4-7 mm broad. *Phoenicaulis*

23b. Plants hairless or pubescent, but not glandular; siliques mostly less than 4 mm wide. **(24)**

24a. Leaves compound, or if simple, plants usually less than 10 cm tall (see *C. bellidifolia*); flowers white, pink, or purplish, or varicolored in populations. *Cardamine*

24b. Leaves simple, often entire (basal one pinnatifid in *Arabis*); plants often over 10 cm tall, or if less than fruit only 8-12 mm long or partition imperfect or lacking; flowers usually white. **(25)**

25a. Flowers in axils of stem leaves (bracts); stems 2-5 cm long, pubescent; partition lacking . *Aphragmus*

25b. Flowers in bractless racemes; stems often more than 5 cm long; hairless partition present or absent;   siliques hairless; leaves mostly on stems. *Arabis*

57

Aphragmus eschscholtzianus   Aphragmus

Rhizomatous herb, stems 2-6 cm long, short, simple hairs or hairless. Leaves mainly basal, entire margin, ovate to flattened spoon-shaped, tapering to base, not clasping. Flowers in short, almost umbel-like racemes, subtended by leaf-like bracts, 4 deciduous sepals, 4 petals, white or purplish. Fruit silique, oblong to ellipsoid, slightly pressed, valves obscurely 1-nerved, hairless.

Southern Alaska. Wet areas in tundra. Uncommon.

Arabis hirsuta   Hairy Rockcress

Herb 20-60 cm tall, stem simple or few to many from taproot, with simple or forked long white hairs at least near base, whole plant sometimes reddish. Basal leaves toothed to entire (1.5-9 cm long, 0.5-2.5 cm broad), oblanceolate, tapering to broad petiole with simple forked hairs, often purplish below, stem leaves comparatively wide spaced, several, toothed, with ear-like appendages at stem (1-5 cm long, 0.3-2 cm broad). Flower in racemes several to many-flowered, pedicels erect, outer sepals pouch-shaped, greenish, petals white or pinkish. Fruit mature siliques erect to spreading, hairless, seeds winged or wingless.

Throughout Alaska south of Brooks Range. Open areas, gravel, rocky slopes, scree, talus, rocky beaches, and roads low to mid elevation. Common. Sometimes infected by a rust fungus that resembles flowers, these bright yellow diseased leaves attract flies which spread the fungal spores. Flowers May-July.

Arabis lyrata   Lyre-leaved Rockcress

Stem more or less branched from basal rosette of leaves, hairless or straight scattered hairs. Basal leaves lyre-shaped to pinnatifid, or oblanceolate, tapering to broad petiole, hairless or hairy, reddish-purple; stem leaves flattened spoon-shaped or oblanceolate, margin entire, toothed or lobed, hairless, dull-purple, stalkless or tapering to broad petiole, bracts in lower leaf axils. Racemes few to many-flowered, two pair of broad opposite petals; sepals greenish or pinkish, hairless or hairy, petals white, aging to pinkish. Siliques linear, hairless, compressed, seeds in one row, not winged.

Most of Alaska. Moist, stony places, scree slopes, roadsides. Common. Flowers May-July. Young leaves have radish flavor, eaten raw or boiled.

Arabis drummondii (Drummond Rockcress) Usually hairless throughout, underside of leaves purplish, petals white or pinkish; seeds broadly winged. Locally common in widely disjunct populations on disturbed soils along streams and roadsides in southern Alaska. Flowers June-July.

Arabis divaricarpa (Spreadingpod Rockcress) 20-70 cm tall from taproot, simple or branched stem, hairless near base or with flat hairs, hairless above; basal leaves entire or toothed (1.2-4 cm long, 0.3-0.8 cm broad), oblanceolate, tapering to a slender petiole, stem leaves oblong to lance-shaped, usually entire margined, stalkless, with earlike appendages, upper ones usually hairless, lower ones mostly hairy; racemes many-flowered, pedicels erect, not subtended by bracts, 4 pink or purplish petals and 4 green sepals; siliques hairless, seeds often in two rows, winged. Dry slopes in disturbed soils from Haines and Skagway area and continental southeast quarter of Alaska. Flowers June-July.

Barbarea orthoceras   **Wintercress**

Hairless herb to 80 cm tall, stem angled, stiff, erect, mostly single. Leaves large, rounded, irregularly toothed to entire, pinnate basal leaves with terminal lobe and 1-2 pair of linear leaflets, upper pinnate leaves clasp stout succulent stem. Dense raceme of flowers becoming loose spike of numerous seedpods, sepals yellowish-green, petals spatula-shaped, petals yellow to pale yellow. Fruits erect siliques lying flat with thick beak, seeds in one row.

Southeastern Alaska west to Aleutians and north to Alaska Range. Banks of streams, wet areas, roadsides, and beaches. Occasional to common. Flowers June-July. Rosettes of dark green shiny leaves cooked as a green vegetable or eaten raw in a mixed salad

Cakile edentula   **Sea Rocket**

Fleshy, smooth, annual herb with stems 10-40 cm tall, branched from base, prostrate or erect. Leaves oblanceolate to oblong with deep wavy margins and stout pedicels. Flowers in racemes, pedicels not subtended by bracts, sepals greenish or pinkish, hairy apically, slight pouch at base, petals white or pinkish. Siliques ribbed and flattened, upper segment tapering to a flattened beak. achenes 2-nerved and hairy.

Islands of southeastern Alaska, west along the coast to Kodiak Island. Dry open slopes, gravel bars, sandy beaches and bogs in Glacier Bay. Flowers in July.

Cardamine oligosperma   Little Western Bittercress

Herb 10-45 cm tall, stem hairless from a slender rhizome. Leaves in a loose basal rosette with 5-9 pinnately divided, rounded leaflets with a large 3-lobed ultimate section, upper stem leaves with lance-shaped or oblong leaflets. Flowers crowded at top of stem, sepals hairless, greenish or purplish, petals white. Fruit linear, erect; siliques hairless.

Southeastern Alaska west to Aleutians. Wet open and disturbed areas. Common. Flowers May-August. Also known as *C. umbellata* and in southeastern Alaska apparently hybridizes with *C. pensylvanica*. *C. pratensis* var. *angustifolium* reported from northern southeastern Alaska.

Cardamine bellidifolia (Alpine Bittercress) Usually less than 10 cm tall, with simple ovate basal leaves and much reduced stem leaves; sepals pinkish, petals white. Arctic and alpine tundra in most of Alaska. Flowers July-August.

Cardamine breweri (Brewer's Bittercress) 20-60 cm tall, mostly stem leaves with 3-5(7) leaflets, terminal leaflet heart-shaped and hairless, sepals green to pink and petals white. Woods of southeastern Alaska. Flowers May-August.

Cardamine occidentalis (Western Bittercress) Perennial herb 20-40 cm tall from tuber-bearing rhizome, basal leaves with 5-7 leaflets, terminal one oval to heart-shaped, often shallowly lobed, upper leaves reduced. Flowers white, 4 petals. Along edges of flowing or standing water. Reported in southeastern Alaska. Flowers May-July.

Cardamine pensylvanica (Pennsylvania Bittercress) 15-35 cm tall, 7-11 oval to lance-shaped basal leaflets, terminal leaflet orbicular, stem leaves reduced upwards, sepals pinkish, petals white. Moist woods, lake shores, and seeps in southeastern Alaska. Flowers May-July.

Cochlearia officinalis   Scurvy-grass

Herb 5-36 cm tall with coarse, fleshy, basal leaves soon withering. Basal leaves ovate or oblong, entire to toothed margin, stem leaves reduced upwards, becoming stalkless. Racemes in dense or compact heads, pedicels not subtended by bracts, tips tinged red, sepals greenish, tinged with yellow or purple, petals white. Silicles round in elongated racemes.

Coastal Alaska. Intertidal to beachgrass zone and on rocks along coast. Common. Flowers late May-August. Leaves eaten raw or boiled; prevents scurvy. Also known as *C. groenlandica.*

KEY TO SPECIES OF *DRABA:*

1a. Basal leaves fleshy, 1-22 cm long, 0.3-4 cm broad, wedge-shaped-oblanceolate, commonly toothed; plants maritime.
D. hyperborea

1b. Basal leaves seldom fleshy, 0.3-5 cm long, 0.1-2 cm broad, oblanceolate to oblong or elliptic; plants maritime or not. **(2)**

2a. Plants with 1-many stem leaves. **(3)**

2b. Plants with a leafless peduncle, lacking stem leaves (or rarely with 1). **(7)**

   3a. Plants annual. **(4)**

   3b. Plants biennial or perennial. **(5)**

      4a. Pedicels usually over 1.5 times longer than silicles; silicles obtuse to rounded apically.
D. nemorosa

      4b. Pedicels usually less than 1.5 times longer than silicles; silicles acute apically.   *D. stenoloba*

   5a. Flowers yellow (occasionally fading whitish).   *D. aurea*

   5b. Flowers white (sometimes cream to yellowish in *D. glabella*). **(6)**

   6a. Lower part of stem with long, spreading, simple hairs, sometimes mixed with forked hairs; plants of coasts and islands of southern Alaska (less common in interior southern Alaska).   *D. borealis*

   6b. Lower part of stem with chiefly starlike hairs, rarely with a few long simple ones; plants of western, northern, and interior Alaska, rarely in coastal southern Alaska.
D. glabella

7a. Flowers white.   *D. lactea*

7b. Flowers yellow (sometimes fading whitish) or purplish.
D. oligosperma

Draba aurea   Golden Rockcress

Herb 10-40 cm tall with tall, straight stem. Basal leaves oblanceolate, somewhat acute, often red below, densely covered on both sides with simple, forked and star-like hairs. Flowers solitary or in lateral racemes in axils of upper leaves, petals rounded, pale yellow. Silicles often twisted and pubescent.

Southeastern Alaska west to Alaska Peninsula. Gravelly soils and rock outcrops. Occasional. Flowers June-July. Highly variable species.

Draba stenoloba   Slender Draba

Stem with star-like hairs, from slender taproot. Rosette basal leaves oblanceolate, toothed or entire with star-like hairs, stem leaves 1-8, ovate to lance-shaped, toothed or entire, hairy. Racemes 3-many-flowered, petals yellow to cream, turning reddish when dried. Siliques linear.

Juneau north and west to Alaska Peninsula. Dry mountain slopes. Occasional. Flowers June-July. Complex genus; hybridization common making species distinctions difficult to impossible.

Draba borealis (Northern Rockcress) Loosely tufted; basal leaves oblanceolate, toothed or entire, stem leaves ovate to lance-shaped, elliptic or oblong, toothed or entire, hairy with 4-6-rayed starlike hairs and often straight simple or forked hairs as well; stems hairy; racemes several- to many-flowered, sometimes with solitary flowers in upper leaf axils, sepals with soft hairs, petals white or cream. Alpine tundra and open woods in southern Alaska. Flowers June-July. Closely related to ***D. glabella*** from which it cannot always be distinguished.

Draba glabella (Smooth Draba) Loosely tufted with oblong basal leaves, entire or with coarse teeth, pubescent, stem leaves have margins with sharp teeth; petals white, notched at apex; silicles hairless. Common in tundra or less common in open woods in much of Alaska except southwestern portion. Also known as ***D. hirta***.

Draba hyperborea (Sub-boreal Draba) Loosely tufted, leaves with short, branched hairs, basal leaves large, tongue-shaped to obovate with large teeth and long petioles, upper leaves entire; inflorescence many-flowered, petals pale yellow; silicles dark brown, often inflated. Coast on cliffs in southern and southeastern Alaska and islands of Bering Sea.

Draba lactea (Tundra Draba) Loosely tufted with lance-shaped leaves having simple or forked hairs in margin of lower part with fine star-like hairs on other half, flowering stem hairless or pubescent, leafless or rarely with one leaflet; inflorescence dense, elongated in fruit, sepals with sparse soft hairs; silicles ovate-lance-shaped, hairless. Tundra, mostly near coast in northern, northwestern, western and disjunctly from alpine sites in southern Alaska. Flowers June-July. Includes *D. nivalis* and *D. lonchocarpa*.

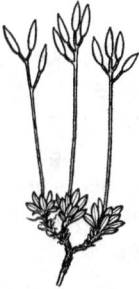

Draba nemorosa (Woods Draba) Annual, stems pubescent, leaves oblanceolate to ovate or oblong, pubescent, pedicels hairless; petals yellow to white; silicles hairless. Weedy species of disturbed or open sites from widely scattered places in southern Alaska.

Erysimum cheiranthoides Wormseed Mustard

Annual herb 20-100 cm tall with sparse, flat lying hairs. Leaves linear to oblong, entire or finely toothed, tapering to base. Flower racemes bractless, sepals yellowish or greenish, 2-3 mm long, petals 3-5 mm long, yellow. Siliques almost circular in transverse section.

Much of Alaska. Weedy introduced species of disturbed sites. More common in Interior Alaska. Flowers June-August. This genus used for poultices in the past.

63

Lepidium densiflorum   Prairie Pepper-grass

Finely hairy annual herb with stems to 50 cm tall, usually branched. Basal leaves in a rosette entire to toothed, usually absent at flowering, stem leaves reduced upwards. Sepals hairy, petals 4, lacking, or reduced, white. Silicles oblong-obovate with shallow apical notch.

Much of southern Alaska. Waste places, beaches, introduced. Uncommon. Flowers June-August.

Parrya nudicaulis   Daggerpod

Branches of woody base with persistent leaf bases, stems 5-30 cm tall, hairy, with slender, stalked to glandular hairs. Leaf margins entire or with sharp pointed teeth, or lobed to pinnatifid. Sepals often tinged pinkish or purplish, petals pinkish, purplish or white. Silique constricted between seeds.

Most of Alaska. Arctic and alpine tundra. Occasional. Flowers May-June.   Also known as *Phoenicaulis nudicaulis*.

Rorippa palustris   Marsh Yellow Cress

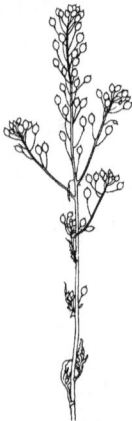

Herb with several stems from thick whitish taproot, stems with coarse hairs or hairless, lower stem and leaves reddish. Lower leaves clasping stem, stalked, lyreshaped-pinnatifid with large almost entire terminal lobe, upper leaves reduced to almost linear lobe with a few teeth at apex. Flowers appear tubular, petals three times as long as yellow sepals, petals yellow. Fruit sausage-shaped, 3-12 mm long, 2-3 mm broad, 2-4 carpelled, styles to 1 mm long.

Most of Alaska. Forest edges, wet open areas, along slow moving water. Occasional to common. Flowers late May-early August. Includes *R. hispida* and *R. islandica.* Also known as *R. barbareifolia*.

Rorripa curvisiliqua (Western Yellow Cress)
Smaller and more creeping than **R. *palustris*** to 40 cm
tall. Leaves oblong to oblanceolate, toothed, lobed or
pinnatifid, stem leaves reduced upwards, clasping stem
with yellow flowers and curved siliques on short stalks.
Disturbed places, open woods in southern Alaska.
Uncommon. Flowers June-August.

Rorripa obtusa (Blunt-leaved Yellow Cress) Hairless
with pinnately divided leaves and small yellowish flowers.
Found rarely in wet places in southeastern Alaska. Flowers
July-August.

Subularia aquatica   Awlwort

Submerged aquatic annual hairless herb with
white fibrous roots. Leaves basal, grass-like to awl-
shaped, 1-5 cm long, tapering to apex. Flowers in
racemes, very small and inconspicuous, 4 white
petals and 4 sepals. Siliques hairless, slightly
flattened, style obsolete.

Most common on islands and along coasts in
southern Alaska, shallow, fine gravelly ponds and
streams; also interior southern Alaska. Occasional
to common. Flowers June-July.

Thlaspi arcticum   Arctic Pennycress

Simple hairless perennial to 20 cm tall. Basal
leaves oblanceolate to obovate, untoothed margins;
stem leaves reduced upwards. Petals twice as long
as sepals, white or slightly lilac. Silicles club-shaped
with thread-like style about 1 mm long.

Known from widely disjunct localities in
southeastern and northeastern Alaska. Arctic and
alpine tundra. Rare, only single specimens collected.
Had been proposed for listing as threatened and
endangered, but has been found to be more
abundant and widespread than originally believed.
Also known as ***Noccaea arctica***.

CALLITRICHACEAE
WATER-STARWORT FAMILY

Callitriche verna   Spring Water-starwort

Amphibious or aquatic herb with more or less prostrate stems floating in water or stranded on mud. Lower leaves linear, 1-veined, upper leaves develop later and are broadly spoon-shaped to ovate, 1 to 3-veined, eventually forming floating rosettes. Flowers inconspicuous, subtended by hornlike bracts, perianth lacking. Fruit stalkless, tan to pale red brown, with pit-like depressions arranged in rows.

Southern half of Alaska, low to mid elevation. Mud, shallow lakes and ponds. Common. Flowers July-August. Also known as *Callitriche palustris*.

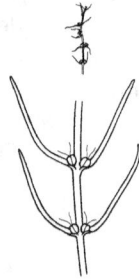

Callitriche anceps (Two-edged Water-starwort) Leaves crowded, flowers not subtended by bracts; brownish fruit scarcely or not at all winged, pit-like depressions not in definite rows. Shallow ponds in southern and western Alaska. Flowers August-September. Also known as *C. heterophylla*.

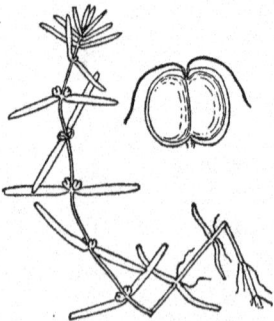

Callitriche hermaphroditica (Autumnal Water-starwort) Leaf tips 2-lobed, flowers not subtended by bracts; grayish fruits narrowly winged, with pit-like depressions not in definite rows. Seldom in flower. Shallow ponds in widely scattered locations in Alaska. Flowers July-September.

CAMPANULACEAE
HAREBELL FAMILY

Campanula lasiocarpa Mountain Harebell

Herb less than 10 cm tall from thin branching rhizome with 1-several stems ascending to erect, hairless or slightly hairy, especially upwards. Leaves in basal rosette, elliptic to oblanceolate, long-petiolated, margin entire or with coarse, sharp teeth; stem leaves stalkless, gradually reduced. Flowers single, large bell-shaped flower per stalk, single calyx with white hairs, calyx lobes sharp pointed with sharp segments, petals blue or rarely white. Fruit oblong, pubescent capsule.

Much of Alaska. Alpine, rocky areas. Occasional. Flowers July-August.

Campanula rotundifolia Bluebells of Scotland

Stems hairless or essentially so 10-80 cm tall, from slender, branched woody rhizome with 1-several prostrate to ascending or erect stems. Mainly stem leaves, basal leaves petiolated, ovate to rounded heart-shaped, angular toothed, soon withering, upper leaves reduced, linear or narrowly lance-shaped, margins almost entire to entire, rough to touch. Flowers solitary, calyx lobes awl-shaped, entire, sepals reflexed, long tapering to a sharp point, purplish-blue, rarely white. Fruit a nodding capsule, oblong, opening through pores at base.

Southeastern Alaska west to Kenai Peninsula and Kodiak Island. Open areas, grassy slopes, and rock outcrops along coast. Occasional to common. Flowers late June-September. Highly variable. Includes *C. latisepala*.

Campanula scouleri　　Scouler's Harebell

Perennial herb 10-40 cm tall from slender rhizomes, stems 1-several, prostrate, ascending or erect, hairless to short-hairy. Mainly stem leaves, lower leaves petiolated, ovate to rounded, sharply toothed, reduced to linear bracts in inflorescence. Flowers on elongate stalk, corolla with ovate-oblong spreading or backward bending lobes about as long as or longer than tube, styles protrude beyond petals, petals pale blue. Fruit an almost spherical capsule.

Southern southeastern Alaska. Forests and rock outcrops, low to middle elevation. Uncommon. Flowers June-August.

Campanula uniflora　　Arctic Harebell

Perennial herb 5-15 cm tall from thick taproot, woody base, stem simple, often prostrate at base, more or less hairy in decurrent lines. Leaves hairless, almost leathery, elliptic to oblanceolate to linear, entire to round toothed, upper leaves hairy, margins rough to touch. Flowers solitary, pedicels hairless or nearly so, calyx lobes lance-attenuate, entire, calyx and ovary hairy, corolla lobes ovate, erect, about as long as tube, styles shorter than corolla, petals blue to purplish. Capsules almost cylindrical, elongate, tapering to base.

Widely disjunct locations in much of Alaska. Rocky alpine areas. Scattered occurrences. Flowers July-August.

CAPRIFOLIACEAE
HONEYSUCKLE FAMILY

Linnaea borealis   Twinflower

Somewhat woody stems less than 10 cm long, trailing, becoming hairless in age. Leaves leathery, evergreen, broadly ovate to roundish, mostly with 1-3 acute teeth on each side, sparsely hairy on margins. Flowering stems erect to ascending with 2(4) pairs of leaves, two nodding, bell-shaped flowers on each stalk, hairy within, petals pink or pinkish. Fruit dry, 1-seeded, with sticky, glandular hairs.

Much of Alaska. Forest, forest edge, bogs and meadows. More common in southern southeastern and interior Alaska. Flowers June-July.

Lonicera involucrata Bearberry Honeysuckle

Deciduous shrub 1-3 m tall, twigs 4-angled when young, hairless, ringed at nodes. Leaves opposite, short-stalked, elliptic, pointed at both ends, edges hairy and not toothed. Flowers paired above, 4 leaf-like green or purple bracts at base of leaves, tubular calyx, funnel-shaped corolla swollen on one side of base, 5 short lobes, 5 hairy stamens, petals yellow. Fruit paired above 4 dark red bracts, few seeded shiny dark berries.

Southern southeastern Alaska, outer coasts on Dall and Prince of Wales Island, also in isolated pockets in Prince William Sound (Kayak Island) and SE Lynn Canal area. Moist soil in open and beach forests. Uncommon and localized. Flowers May-August. Bitter fruits said to be poisonous. Bark and twigs used to treat digestive tract problems and as a contraceptive. Purple juice from berries used as a pigment to dye roots for baskets. Haida rubbed berries into scalp to prevent hair from turning gray.

Sambucus racemosa **Pacific Red Elderberry**

Clump-forming shrub or small tree to 6 m tall with strong odor. Leaves opposite, pinnately compound, leaflets 5-7, opposite except at end, leaflets lance-shaped to egg-shaped, sharply toothed. Flowers in terminal clusters with many small ill-scented flowers that turn brown when dried, petals whitish. Fruit bright red or purple-black, or less often yellow, brown or sometimes white berry with 2-5 seeds per berry.

Southeastern Alaska west to Alaska Peninsula, less common in Interior. Meadows, openings in forests, in brushy thickets with alder or willow, along streams. Common. Flowers May-June; fruits July-August. Raw berries cause nausea; stems, bark, leaves and roots contain cyanide-producing glycosides that cause diarrhea and vomiting. Berries strained of seeds edible but have unpleasant taste. Tea made from inner bark used for colds, influenza, and tuberculosis; and stem bark salve used externally as wash for infections. Seeds persist in soil and germinate when soil is disturbed by windthrow, burning, etc. Also known as **S. callicarpa**.

Symphoricarpos albus **Snowberry**

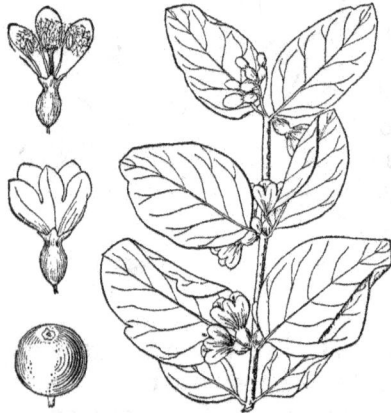

Deciduous much-branched shrub to 2 m tall. Leaves opposite with slender petioles, thin, elliptic to ovate with a few irregular teeth. Flowers few in short clusters at ends of twigs or bases of upper leaves, tubular corolla, petals pink. Fruit round white berries.

Sitka, north of Juneau to head of Lynn Canal including along the river in the Chilkat Bald Eagle Preserve. Forest edges, rocky alpine meadows; sometimes introduced in towns. Occasional. Flowers June-July. While eaten by grouse and other birds, the berries considered poisonous to humans and contain the isoquinoline alkaloid chelidonine, and other alkaloids. Ingesting berries causes mild symptoms of vomiting, dizziness and slight sedation in children

Viburnum edule Highbush Cranberry

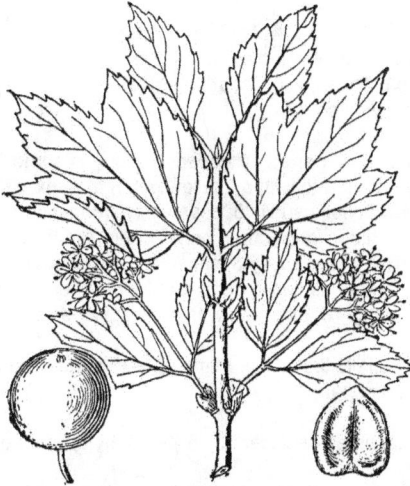

Deciduous shrub 0.5-3.5 m tall with smooth gray bark and thick twigs with rings at nodes and white pith. Leaves opposite, shallowly and palmately three-lobed, three main veins from rounded base, edges sharply toothed and lobes short-pointed. Flowers in terminal clusters on short lateral twigs with several to many short-stalked flowers, petals whitish. Fruit elliptic, fleshy, non splitting, red or orange containing 1 rounded flat pit.

Most of Alaska except Aleutians and extreme northern coastal plain. Forest edge, beaches, bogs, tidal meadows and along streams. Common. Flowers May-July; fruits July-September. Fruit edible raw and makes good jelly when picked not quite ripe; makes excellent wine. Wildlife eat leaves and berries. Bark natural source of muscle relaxant: One cup of bark tea made from a handful of bark shavings stops menstrual cramps. Boil and drink tea for colds, sore throat and laryngitis.

CARYOPHYLLACEAE
PINK FAMILY

KEY TO GENERA OF *CARYOPHYLLACEAE:*
1a. Leaves with distinctive thin, dry, translucent stipules.   *Spergularia*
1b. Leaves without thin, dry, translucent stipules. **(2)**
2a. Sepals distinct or nearly so. **(3)**
2b. Sepals united, forming a tube. **(6)**
    3a. Petals 2-cleft, -parted, or -lobed, (lacking in some *Stellaria* species). **(4)**
    3b. Petals rounded to with a shallow notch at apex. **(5)**
        4a. Styles usually 5; capsule cylindrical, membranous, often 1-2 times longer than calyx, dehiscent by 10 teeth.   *Cerastium*
        4b. Styles usually 3, capsule ovoid to short-cylindrical, seldom as long as calyx, dehiscent by 6-8 teeth.   *Stellaria*
    5a. Styles usually 5, alternate with sepals; valves of capsule opposite sepals.   *Sagina*
    5b. Styles usually 3, if more, than opposite sepals; valves of capsule alternate with sepals.   *Arenaria*

6a. Styles usually 5; capsules dehiscent by 5 (or 10, by splitting of primary) valves.
   Lychnis
6b. Styles usually 3; capsules dehiscent by 6 (or rarely by 8-10) valves.   *Silene*

KEY TO SPECIES OF *ARENARIA:*
1a. Plants with fleshy stems and leaves; flowers more or less unisexual; plants of seashores.   *Honkenya peploides*
1b. Plants not at all fleshy, or if slightly so, then plants with both pistils and stamens and seldom maritime. **(2)**
2a. Leaves ovate to lance-shaped, oblanceolate or broadly elliptical, acute to rounded apically. **(3)**
2b. Leaves linear to oblong or awl-shaped, acute, abruptly acute, or blunt apically. **(4)**

    3a. Sepals broadly spreading or ascending in flower, usually tinged purplish; capsule inflated at maturity, 7-10 mm long.   *A. physodes*
    3b. Sepals erect (seldom spreading) in flower, usually greenish; capsule not inflated at maturity, 5 mm long or less.   *A. lateriflora*

4a. Sepals acute to acuminate; plants slender, annual, biennial, or short-lived perennials from weak taproots.   *A. rubella*
4b. Sepals broadly rounded at end; plants slender to coarse perennials with strong taproots. **(5)**

5a. Leaves more or less flattened, strongly ciliate marginally, 3-veined; capsules mostly 10-15 mm long.   *A. macrocarpa*
5b. Leaves seldom flattened; capsules 10 mm long or less.   *A. sajanensis*

Arenaria lateriflora   Blunt-leaved Sandwort

Herb 1-4 cm tall with threadlike, branched, more or less horizontal rhizome, flowering stems erect, thin, usually branched. Leaves blunt or rounded at tip, basal leaves smaller than upper ones, largest near middle of stem, lance-like to elliptic, pinnately three-veined, minutely hairy below, especially along midrib. Flowers erect, solitary or 2-5 in terminal cymes, sepals with whitish, thin membranous margins, 5 obovate white petals two to three times as long as 5 sepals. Capsules 6-valved, ovate, twice as long as sepals, seeds almost kidney-shaped, black, shining.

Most of Alaska. Dry meadows, forest edges and open places, river flats and sandy beaches low to mid elevation. Common. Flowers June-August. Also known as *Moehringia lateriflora*.

Arenaria physodes   Merckia

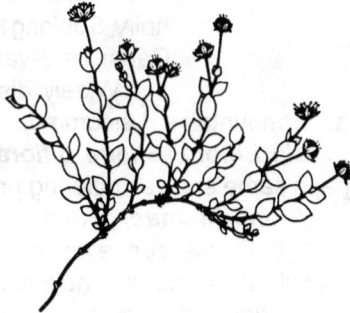

Stem creeping, strongly branched, more or less mat-forming, flowering stems prostrate to erect. Leaves ovate to oblong, acute, hairless except margin fringed with hairs, lower leaves smaller than those at mid stem, stem leaves 3-7 pairs or more, lance-shaped to ovate, more or less fleshy, hairy below especially on mid vein or hairless, sometimes with sterile axillary shoots. Flowers erect, solitary, terminal or axillary, pedicels hairy, sepals with whitish or purplish membranous margins, 5 oblanceolate white petals. Capsules almost spherical, 6-lobed, 6-valved, seeds roughened, yellowish brown.

Most of Alaska. Sandy areas, river banks. Occasional. Flowers June-July. Also known as *Wilhelmsia physodes*.

73

Arenaria rubella   Reddish Sandwort

Forms cushions or mats arising from taproot and branching woody base with persistent leaves, flowering stems erect or ascending. Leaves basal lance-linear, 3-ribbed, acute apically; stem leaves 2-6 pairs, similar to basal ones but reduced upwards, often with short sterile axillary branches. Flowers small, erect, solitary, or more commonly 2-4, stalks glandular-hairy, sepals commonly reddish-purple, acute apically, petals oblong-lance-like, white. Capsules 3-4 valved, egg-shaped, seeds reddish-brown, minutely roughened.

Most of Alaska except Aleutians. Rocky, sandy areas in alpine, open woods and roadsides. Occasional to common. Flowers June-August. Also known as *Minuartia rubella*.

Arenaria sajanensis   Sajan Sandwort

Perennial herb forming dense to moderately dense tufts arising from taproot and branching woody base with persistent leaves, flower stems erect or nearly so. Basal leaves linear, 1-veined, hairless or margins fringed with hairs in lower portion; stem leaves (0)2-3 pairs, often broader than basal ones. Flowers erect, solitary or in pairs, pedicels glandular-hairy, sepals sometimes purplish at tips, glandular-hairy, 5 oblong to narrowly oblanceolate white petals. Capsules 3-valved, seeds reddish-brown, smooth.

Widely disjunct throughout Alaska. Dry open areas in mountains. Uncommon. Flowers June-July. Also known as *Minuartia biflora, Silene biflora. A. obtusiloba* closely related or perhaps same species differing primarily in its larger size.

Arenaria macrocarpa (Long-podded Sandwort) Forms dense mats 5-40 cm broad, base clothed with persistent leaves, basal leaves lance-awlshaped to lance-oblong, 3-veined, margin fringed with hairs, stem leaves in 1-6 pairs, often broader than basal leaves; flowers erect, solitary, pedicels glandular-pubescent with multicellular hairs, sepals often purplish, glandular hairy to hairless in age; petals white, obovate; seeds almost kidney-shaped, strongly warty. Arctic and alpine tundra in most of Alaska but primarily Interior. Flowers June-July. Also known as *Minuartia macrocarpa*.

Cerastium arvense Field Chickweed

Grayish-green, finely hairy, tufted, stem branched from base with sterile axillary shoots 5-50 cm tall, glandular-hairy on upper stems and flowers, highly variable in hairiness and leaf form. Leaves opposite, linear, lower persist after withering, stem leaves with sterile axillary shoots. Flowers with 5 fairly showy white petals 2-3 times as long as sepals, deeply notched. Capsule cylindric, seeds golden to reddish-brown opening by ten teeth.

Southern third of Alaska. Meadows, dry slopes and roadsides. Common. Flowers May-July.

Cerastium earlei Bering Chickweed

Matted perennial, stem leaves lacking sterile axillary shoots or bearing them only in lowermost axils, densely hairy becoming glandular above. Basal and stem leaves similar, narrowly oblong to spatula-shaped. Flowers fewer than *C. arvense* with 5 deeply bilobed white petals longer than calyx, sepals often purplish. Cylindric capsules much longer than calyx.

Juneau north throughout Alaska. Arctic and alpine tundra. Occasional. Flowers June-August. Hybridizes with *C. arvense*. Also known as *C. beeringianum*.

Honkenya peploides Beach Greens

Fleshy herb forming loose to dense mats, hairless, from taproot buried in sand. Leaves yellowish-green, ovate, oblong or lance-shaped, stem leaves 3-10 pairs or more. Flowers erect, solitary, scattered in upper leaf axils, sepals spreading in flower, greenish, hairless, acute apically, 5 obovate white or greenish-white petals. Capsules almost spherical, 3-5-valved, seeds brown, smooth shiny.

Coastal Alaska. Sandy beaches just above mean high tide. Common. Flowers June-July. Also known as *Arenaria peploides*. Mixed with fat and berries to make "Eskimo ice cream"; also used to make kraut and cure scurvy. When eaten raw, tastes like green beans.

Lychnis furcata   Arctic Lychnis

Hairy herb to 20 cm or taller. Basal leaves lance- to spatula-shaped, hairy, with small light colored hairs especially in margins; stem leaves in 1 or 2 pairs, oblong, shorter than basal leaves. Flowers erect, solitary, or sometimes 2-3, densely hairy, inflated in age, calyx bell-shaped to urn-shaped, 5 bilobed white to pinkish petals. Capsule with narrowly winged margin.

Most of Alaska except southwestern portion. Dry places, meadows, alpine. Uncommon. Flowers June-July. Also known as *Melandrium affine* and *Silene involucrata.*

Lychnis apetala (Nodding Campion) To 30 cm tall with nodding flowers in bud, becoming 5 pinkish, purplish, or white pinwheel-like petals; seeds with a broad, more or less inflated winged margin. Uncommon. Meadows, rock outcrops, and lake shores in much of Alaska from Juneau north. Flowers June-July. Also known as *Melandrium apetalum*. This species has been split into many species mostly in the *Silene* genus.

Sagina crassicaulis   Coastal Pearlwort

Herb forming dense, dark-green tufts to 20 cm tall with broader leaves in center. Basal leaves in rosette, linear, 1-veined, sharp-pointed; stem leaves 2-6 pair usually with clustered secondary axillary leaves. Flowering stems prostrate to erect, flowers erect, solitary, terminal or lateral, sepals with purple spots in wax-papery margins, petals white, 4 (5 or lacking) shorter than sepals. Capsules 4-5-valved, erect at maturity, seeds reddish brown.

Southeastern Alaska west to Aleutians. Beaches. Common. Flowers June-July. This species also known as *S. maxima*. Sagina means fodder, since it was used as animal feed.

Sagina intermedia — Snow Pearlwort

Herb forming dense, dark green tufts with broader leaves in center, stems never rooting at nodes. Basal leaves linear; stem leaves 1-8 pairs usually with clustered axillary leaves. Flowers erect, solitary and terminal, or axillary, sepals with purple margin or purplish spots in margin, usually 4, petals white or lacking, shorter than sepals. Capsules erect at maturity, seeds reddish brown.

Juneau north and west to Aleutians. Sandy places, snow beds, rock outcrops and avalanche slopes. Occasional. Forms hybrids with *S. crassicaulis*. Also known as *S. nivalis*.

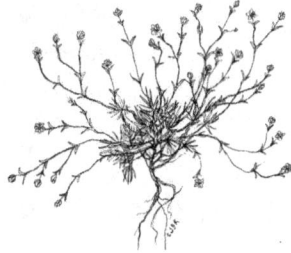

Sagina occidentalis (Western Pearlwort) Annual without a basal rosette and without clustered secondary leaves in axils, usually several flowers to a stem. Uncommon. Moist places in coastal southern Alaska. Flowers June-July. Also known as *S. decumbens*.

Sagina saginoides (Arctic Pearlwort) Forms yellowish-green mats, clustered axillary leaves and thin, dry, translucent sepal margins, not purple, fruit commonly bent downward or backward at maturity. Difficult to distinguish from *S. intermedia* without these characteristics. Moist soil, beaches, mud flats and alpine meadows from southeastern Alaska west to Kenai Peninsula and eastern Aleutians and Interior Alaska. Flowers June-August.

Silene acaulis — Moss Campion

Densely tufted herb 3-6 cm tall, forming dense mats or cushions, stem densely covered with stout flat leaves, margins fringed with stiff hairs. Basal leaves linear to lance-linear 1(3)-veined; stem leaves mostly lacking, reduced pair near stem base. Flowers erect, solitary, stalkless, calyx hairless, mostly reddish, peduncles short, petals pink, lilac or purple, rarely white. Capsule 3-chambered, as long as calyx, sometimes longer, seeds light brown.

Throughout Alaska. Moist but well-drained sites in alpine, sandy soil, lake shores and ridges. Common. Flowers June-August.

77

Silene menziesii (Menzies' Campion) Not densely tufted and taller, generally glandular-hairy on leaves and upper stem, basal leaves well developed, often withered by flowering time, stem leaves opposite and much longer. Flowers white. Open woods and meadows in coastal northern southeastern and Prince William Sound. Occasional. Flowers July-August.

Spergularia canadensis Canada Sand-spurry

Small hairless to glandular-hairy annual 4-30 cm long, forming clumps from a taproot. Leaves opposite, fleshy, blunt, stipules ending abruptly or with a small pointed tip. Flowers in leafy cymes, sepals ovate, mostly longer than petals, petals white to pink, stamens 2-5. Capsules longer than sepals, seeds brown.

Southeastern Alaska west to Kenai Peninsula and Kodiak Island. Saline soil, beaches, salt marshes and on rocks. Occasional. Flowers late June-July.

KEY TO SPECIES OF *STELLARIA:*

1a. Stems distinctly covered with long, soft, somewhat wavy hairs; sepals often fringed marginally with numerous, short hairs. *S. longipes*

1b. Stems hairless or only sparingly covered with long, soft, somewhat wavy hairs; sepals not fringed with hairs, or if so, then hairs long, few, and near base only. **(2)**

2a. Leaves 3-8 mm long, 1-2 mm broad; stems 0.4 mm broad or less; sepals 2.5-4 mm long. *S. crassifolia*

2b. Leaves often over 8 mm long or over 2 mm broad or both, or if less (as in some *S. humifusa* and *S. umbellata*), then sepals mostly 4-5 mm long and plants dying brownish or flowers in axils of membranous bracts. **(3)**

3a. Leaves linear or narrowly oblong, minutely toothed marginally (use at least 30X magnification); flowers few to many, borne in open, membranous-bracted cymes. *S. longifolia*

3b. Leaves lance-linear to lance-attenuate, lance-shaped, lance-ovate, or oblong, margin minutely toothed or entire; flowers solitary or few to several in leafy or membranous-bracted cymes. **(4)**

4a. Leaves lance-ovate to lance-shaped, entire, shortly stalked; flowers solitary in leaf axil. *S. crispa*

4b. Leaves lance-attenuate to oblong, or less commonly lance-shaped, stalkless, or if shortly stalked, then margin minutely toothed. **(5)**

5a. Leaf margin minutely toothed (use a strong lens), and often ciliate with multicellular hairs; leaves commonly 3-9 mm broad; flowers axillary, or in leafy bracted cymes. *S. calycantha*

5b. Leaf margin entire; leaves 1-3 mm broad, or if 3-5 mm broad, then flowers usually solitary. **(6)**

6a. Leaves lance-attenuate, commonly 5-10 times longer than broad or more; flowers solitary, or 2-5 in open cymes. *S. longipes*

6b. Leaves lance-shaped, commonly less than 4 times longer than broad; flowers solitary or rarely 2-3. **(7)**

7a. Flowers borne in axils of membranous bracts; sepals (6)7-9 mm long. *S. alaskana*

7b. Flowers borne in axils of foliage leaves; sepals mostly (3.5) 4-6 mm long. **(8)**

8a. Stems elongate, with 1-several flowers along its length; leaves fleshy, not leathery or shining; plants broadly distributed. *S. humifusa*

8b. Stems more or less contracted, commonly 1-flowered; leaves shining, leathery, not or somewhat fleshy; plants found on islands of southeastern Alaska (Hidden Glacier) and Aleutians. *S. ruscifolia*

Stellaria calycantha   Northern Starwort

Herb forming moderate to large clumps, flowering stems 5-50 cm tall, hairless or sparsely hairy. Leaves opposite, lance-shaped to long-tapering, mostly stalkless, hairless, minutely saw-toothed. Flowers solitary and axillary or more commonly in open terminal cymes, sepals 1-3 veined, margins rough textured, not fringed with hairs, petals white or greenish, sometimes reduced or lacking. Capsules straw-colored to purplish, opening by 6 teeth; seeds reddish-brown, smooth to slightly roughened.

Most of Alaska south of Brooks Range. Wet meadows, river banks, thickets, open woods and roadsides. Flowers June-August. Includes *S. sitchana*. Also known as *S. borealis.*

Stellaria crispa   Crisp Starwort

Hairless mat-forming herb with weak stems 5-50 cm long from slender rhizome. Leaves paired, ovate, stalkless, sharp pointed, slightly wavy in margin. Flowers solitary in leaf axils, sepals lance-shaped, rough margined, petals 5, tiny, white, shorter than sepals or lacking, 3 styles. Capsules straw colored or brownish.

Southeastern Alaska west to Alaska Peninsula, tip of Aleutians. Wet rocks, wet soil on stream banks, in shady alder forests and thickets, low to mid elevation. Common. Flowers June-August.

Stellaria humifusa   Salt Marsh Starwort

Hairless, yellow-green herb 5-40 cm tall, branches rooting at lower internodes, forming mats or clumps, can be elongate and spreading. Leaves lance-shaped to elliptic, entire, fleshy. Flowers solitary in leaf axils, sepals 3-veined, rough margined, sepals shorter than petals, 5 petals appearing as 10, white. Capsule with smooth seeds.

Coastal Alaska. Seashores, lake shores and marshes. Common. Flowers late May-July. May be confused with *S. crassifolia*, but has fleshy leaves that wrinkle and tend to turn brownish when dried; sepals generally longer.

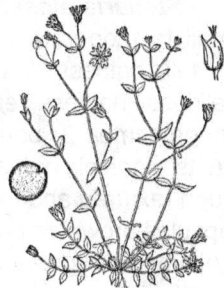

79

Stellaria longifolia　　Longleaved Starwort

Herb forming loose clumps, flowering stem 10-35 cm long, hairless or rough textured. Leaves linear to narrowly elliptic, blue-green, stalkless, hairless or sparingly fringed with hairs, with minute teeth marginally. Flowers few to many in open cymes, rough textured bracts, sepals obscurely 3-veined, membranous-margined, margins not fringed with hairs, white. Capsules greenish or yellowish, opening by 6 teeth, seeds brown, slightly roughened.

Southeastern quarter of Alaska. Wet meadows, bogs and roadsides. Occasional. Flowers June-July. Similar to *S. calycantha*.

Stellaria longipes　　Longstalked Starwort

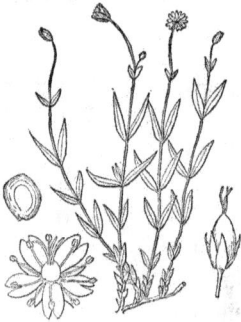

Herb forming small to large clumps or mats 10-40 cm tall with 4-angled stems. Leaves lance-shaped to long-tapering, stalkless, some with long soft wavy hairs, margin entire, fringed with hairs. Flowering stem 3-32 cm long, with long soft wavy hairs, or hairless, flowers solitary or few in open cymes, sepals 3-veined, may be marginally fringed with hairs, white. Capsules often brown or purplish, opening by 6 teeth, seeds brown, slightly roughened.

Most of Alaska. Rocks, stream banks, lakes shores and open woods. Common. Flowers June-July. Includes *S. laeta* and *S. monantha*.

Stellaria alaskana (Alaska Starwort) Forms small to moderate clumps, flowering stem to 30 cm tall, hairless, leaves lance-shaped to elliptic, stalkless, hairless, entire margin; flowers solitary or sometimes 2, borne in axils of rough textured bracts, pedicels hairless, sepals 3-veined, margins rough textured and fringed with hairs, petals white; capsules brown-colored, opening by 6 teeth; seeds light brown, roughened. Rock outcrops, talus slopes, and alpine tundra in interior (Denali National Park and eastward) and southeastern Alaska. Flowers July-August. Some taxonomists include this species with *S. longipes*.

80

Stellaria crassifolia (Fleshy Starwort) Forms small to large mats or clumps, flowering stem 3-30 cm tall, hairless, leaves attenuate to lance-linear, stalkless or nearly so, hairless, entire margin; flowers solitary or few to several in open cymes, sepals 3-veined, margins rough textured, not fringed with hairs, petals white; capsule straw-colored, opening by 6 teeth; seeds reddish brown, roughened. Moist soils in bogs, open woods and lake shores in much of Alaska except Aleutians. Flowers July-August.

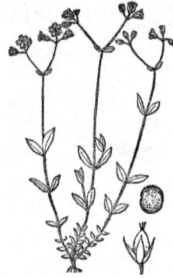

Stellaria ruscifolia (Thick-leaf Starwort) Forms small to moderate clumps, flowering stems 3-12 cm tall, hairless, leaves ovate-lance-shaped to lance-shaped, stalkless, hairless, entire margin; flowers solitary in axils of foliage leaves, pedicels hairless, sepals obscurely 3-veined, petals white; capsules straw-colored, opening by 6 teeth, seeds roughened and brown. Found rarely, gravelly sites on islands of southeastern Alaska and Aleutians. Flowers June-August.

CHENOPODIACEAE
GOOSEFOOT FAMILY

Atriplex patula   Spearscale

More or less mealy herb, 10-100 cm tall, erect or prostrate, stem usually much branched. Lower leaves opposite, upper leaves alternate, stalked to stalkless, leaf margins entire or with sharp teeth from triangular to arrow-shaped to lance-shaped to linear. Flowers inconspicuous in leafy axillary spikes, greenish. Fruit with membranous wall, seed usually erect.

Southeastern Alaska west to western coast, less common in Interior Alaska. Seashores, saline soils. Common. Flowers July-August. Leaves vary; a complex of **A. gmelinii**, **A. alaskensis**, **A. drymarioides**, and **A. patula**. Same family as spinach. Leaves and tips of young plants make excellent salad greens, but use sparingly as it contains oxalate salts.

Salicornia virginica   Sea Asparagus

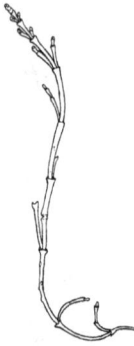

Small and shrubby, to 30 cm long from elongate rhizomes, stem with stout branches, segments longer than broad. Leaves scale-like, often with margins rough to touch. Flowers in numerous spikes, flowering stem brownish-purple, central flower slightly above lateral ones, tips of three flowers form uniform arc, petals lacking. Fruit an utricle enclosed in the spongy calyx, covered with backward or downward facing hairs, splits when ripe.

Maritime southeastern Alaska. Alkaline soils, seashores and tide flats. Occasional. Flowers July-August. Eaten raw in "Hydaburg Salad" following good rinse; excellent cooked, stores well. The ashes from this plant, also known as glasswort, provided alkali used in glass making. Also known as *S. pacifica* and *S. depressa.*

Salicornia europaea (Slender Glasswort) Stems usually upright and much-branched 5-15 cm tall, often turning bright red. Central flowers three at each node and rises above lateral ones and is somewhat smaller. Salty soils on tidal flats and lake shores in coastal southern Alaska. Flowers July-August. Also known as *Salicornia maritima*.

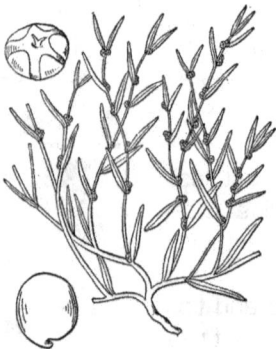

Suaeda maritima   Low Sea-blite

Herb to 30 cm tall from a slender taproot. Leaves linear to oblong, fleshy, somewhat circular in transverse section. Flowers clustered in leaf axils or sometimes terminal and spike-like, 4-5 perianth lobes, fleshy, basally united, petals lacking. Fruit seedy, seeds shiny black, 1-2 mm diameter, beaked.

Southeastern Alaska west to Prince William Sound and Kenai Peninsula. Seashores. Occasional. Flowers July-August. In medieval times, it was harvested and burned and the ashes processed as a source for sodium carbonate for use in glass-making.

CORNACEAE
DOGWOOD FAMILY

Cornus canadensis   Bunchberry

Erect stem 5-25 cm tall from creeping rhizome with 1-2 pair of small opposite leaves and a umbrella-like whorl of 4-6 leaves below flower. Leaf lateral veins arise from mid vein in lower third of leaf. Four, white petal-like bracts surround tiny grouped flowers having 4 green petals, flowers entirely covered with hairs. Fruit red clustered pulpy berries.

Southeastern Alaska west to Aleutians north to 68th parallel. Forest, forest edge, open areas and bogs. Common. Flowers June-July. Fruit edible but insipid and dry; best used in combination with other berries. Tea used to calm colic in infants, leaves as tobacco substitute. Important winter browse for deer.

Cornus suecica (Swedish Dwarf Cornel) Similar to *C. canadensis* but with 2 or more pair of larger stem leaves below whorl, lateral veins arise from base of leaf or nearly so; flower bracts white, yellowish, pinkish, or purplish, petals purplish-black, white hairs at base only; fruit red berries. Bogs, headlands and alpine from southeastern Alaska west to Aleutians and western coast. Forms hybrids with *C. canadensis*.

Cornus sericea   Red-osier Dogwood

Deciduous shrub 1-6 m tall with several stems, bark gray, smooth to slightly furrowed into flat thick plates, twigs dark red, mostly hairy when young. Leaves opposite (paired) with hairy petioles, blades elliptic to ovate, margins entire, 5-7 long curved sunken veins on each side of midrib. Flowers in terminal clusters, flat with many crowded flowers, hairy with 4 minute sepals united at base and 4 white petals, 4 alternate stamens, inferior pistil. Fruit round whitish or light blue with 1 nutlet.

Southeastern Alaska north to southeastern third of Interior Alaska. Moist woods, along streams, rocky areas with seeps and beach forests. Occasional. Flowers June-July; fruits July-September. Dried bark used as tobacco substitute and bark fiber used to make cords for tying and lashing. Also known as *C. stolonifera*. Important winter browse for deer and moose.

83

CRASSULACEAE
STONECROP FAMILY

Crassula aquatica   Water Pygmyweed

Tiny succulent annual herb from taproots, tufted or matted, stem prostrate to erect. Leaves opposite, fleshy, linear to narrowly oblanceolate, flattened, entire margin, with sheaths that join at base. Flowers tiny, borne singly or clustered in leaf axil, 4 sepals connected about half their length, 4 membranous greenish or reddish petals, stamens much shorter than petals. Follicles erect.

Coastal southern Alaska. Inundated shores. Often overlooked due to its small size. Flowers May-July. Also known as *Tillea aquatica.*

Sedum oreganum   Oregon Stonecrop

Tiny succulent annual herb to 15 cm tall. Stems prostrate, base of stem clothed with persistent, scale-like leaves. Leaves egg- to spoon-shaped, reduced and scale-like below, becoming larger upwards, flattened, irregularly toothed to entire, succulent. Stamens longer than petals, filament purple or yellow, usually 5 yellow, aging pinkish petals united near base. Follicles erect, style spreading.

Southeastern Alaska. Rocky ridges and talus slopes, forest edges low to mid elevation. Uncommon. Flowers July-August. Leaves edible.

Sedum roseum   Roseroot

Hairless, succulent perennial to 20 cm tall; rhizome thick, fleshy, scaly, fragrant when cut. Leaves oval to oblong, reduced and scale-like below, becoming larger above, flattened, irregularly toothed to entire. Flowers in terminal head-like cluster of tiny 4-petaled dark wine-colored flowers on fleshy stems. Red or purplish erect follicles with divergent tips.

Most of Alaska. Open rocky areas, coastal rock outcrops, scree slopes and alpine. Common. Flowers late May-August. Leaves and young shoots eaten raw or boiled. Tea used as eyewash or drink for colds, sore throat or mouth, or as a dressing for cuts. Also known as *S. integrifolium* and *Rhodiola rosea.*

Sedum divergens (Spreading Stonecrop) Creeping succulent perennial to 15 cm tall. Leaves of horizontal stems mostly opposite, broadly oval, thick, circular in transverse section, oblong to spoon-shaped on flowering stem; 5 bright yellow petals, oblong-lance-shaped. Rocky cliffs. Found at mile 10 on Haines Highway. Tea made from leaves used during childbirth and breast feeding.

Sedum lanceolatum (Lance-leaved Stonecrop) Hairless, succulent, perennial to 25 cm tall. Lance-shaped to elliptic, fleshy leaves, rounded in cross section; petals 5, yellow, lance-shaped, not united. Open woods and grasslands in southeastern Alaska. Flowers July-August.

DIAPENSIACEAE
DIAPENSIA FAMILY

Diapensia lapponica   Diapensia

Evergreen subshrub forming a mat or tuft 2-8 cm high, stem prostrate and much branched with dense mat of old leaves beneath, twigs slender, hairless, concealed by leaves. Leaves thick and fleshy, crowded and overlapping, narrowly oblong or spoon-shaped, entire margins, edges turned under, wedge-shaped basally. Flowering stems leafless or with a few leaves, flowers solitary, immediately subtended by 3 bracts, corolla bell-shaped with 5 rounded spreading lobes, sepals becoming woody and surrounding fruit at maturity, petals white to cream, rarely pink to red. Fruit ovoid capsules 4-6 mm long with numerous seeds.

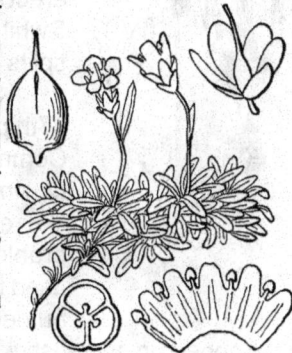

Mountains above Haines and Skagway and nearly throughout Interior Alaska. Rocky alpine areas. Flowers late May-July.

DROSERACEAE
SUNDEW FAMILY

Drosera anglica   Long-leaf Sundew

Insectivorous, stemless perennial 6-18 cm tall. Leaf blades ascending, oblong to linear, tapering at base and two times longer than wide. Flowers fewer and slightly larger than *D. rotundifolia*, scapes straight, erect, 1-several flowers in 1-sided racemes, emerging from center of rosette leaves, 5 white petals and 5 sepals. Capsule 5-7 mm long, splits open on back exposing numerous seeds.

Southeastern Alaska west to Kodiak Island, northern Alaska Peninsula, rarely northward. Peat bogs. Occasional. Flowers June-August.

Drosera rotundifolia   Round-leaf Sundew

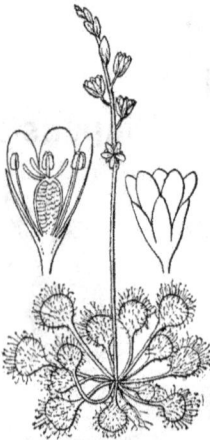

Insectivorous, stemless perennial 6-20 cm tall. Leaves in basal rosette, blades prostrate, oval to round, covered with sticky, red-stalked glands that trap insects, leaves abruptly narrowed into slender petiole with soft hairs. Flowers on slender, erect scape rising above leaves, small flowers with 5 white petals and 5 sepals. Capsule 5-7 mm long, splits open on back exposing numerous seeds.

Southeastern Alaska to Aleutians north to 67th parallel. Peat bogs, swamps, and wet areas. Common. Flowers June-August. Hybrids with *D. anglica* occur. Sap and leaves have ability, like rennet, to curdle milk; tea used for respiratory problems, contains antibiotic substance, also used to treat warts. Sundews capture and absorb nutrients, mainly nitrogen from small insects that get trapped on sweet sticky-haired leaves. The captured mosquitoes, midges, gnats and moths also pollinate the Sundew.

ELAEAGNACEAE
OLEASTER FAMILY

Shepherdia canadensis Soapberry

Deciduous shrub to 2 m tall with silvery or reddish brown minute scales. Leaves opposite, ovate, entire, slightly hairy above, twigs gray, scaly with paired branches. Flowers small in short lateral spikes before leaves emerge, petals yellowish or brownish. Fruit elliptic, red or yellowish, nearly transparent, fleshy.

Most of Alaska except southwestern portion. Open, well drained forests in rocky areas, gravel bars, beach forest. Common (uncommon in southern southeastern Alaska). Flowers May-July; fruits July-August. Fruit bitter, eaten mixed with sugar and water and beaten until frothy for "Indian ice cream". Tea made from stem said to cure tuberculosis or used to wash cuts and swellings.

EMPETRACEAE
CROWBERRY FAMILY

Empetrum nigrum Crowberry

Low, creeping, matted, evergreen subshrub to 20 cm tall with horizontal, much branched stems. Leaves crowded, 4 in a whorl or alternate, long, linear, margins rolled under, spreading. Flowers inconspicuous and solitary in upper axils with 3 bracts, 3 sepals, 3 purplish spreading petals, 3 stamens much longer than petals. Fruit blue black, berry-like with sweet watery juice, 6-9 hard reddish brown seeds.

Widespread in Alaska. Bogs and moist rocky slopes from sea level to alpine. Common. Flowers June; fruits August, persisting under snow throughout winter. Berries eaten raw, cooked, or dried; beer and sparkling wine made from juice. Boiled root and bark said to cure cataracts.

ERICACEAE
HEATHER FAMILY

KEY TO GENERA OF *ERICACEAE:*

1a. Ovary inferior or apparently so. **(2)**

1b. Ovary superior. **(3)**

    2a. Plants evergreen; ovary in reality superior, but surrounded by fleshy, purplish black or whitish calyx when mature.   *Gaultheria*

    2b. Plants mostly deciduous (except for immature, nonflowering forms); ovary truly inferior; fruit red or blue.   *Vaccinium*

  3a. Leaves less than 6 mm long, thick, partly overlapping, usually 4-ranked on stem.   *Cassiope*

  3b. Leaves larger, mostly with well-developed blades, if less than 6 mm long, then not 4-ranked. **(4)**

4a. Corolla with petals distinct or nearly so, not at all tubular. **(5)**

4b. Corolla of united petals, urn shaped, bowl shaped, or bell shaped. **(6)**

    5a. Flowers in terminal corymbs; petals white, less than 8 mm long; leaves evergreen, villous-tomentose with yellowish, reddish, or brownish hairs; margins rolled.   *Ledum*

    5b. Flowers solitary, terminating lateral branchlets (rarely axillary); petals salmon or copper colored, 10-15 mm long; leaves deciduous, hairless beneath, flat.   *Cladothamnus*

6a. Corolla broadly cone-like to saucer shaped or rotate, not constricted at apex. **(7)**

6b. Corolla bell- to urn-shaped, more or less constricted at apex. **(9)**

    7a. Leaf blades 2-8 mm long, distinctly opposite; corolla 4-5 mm long.   *Loiseleuria*

    7b. Leaf blades mostly over 10 mm long, alternate or opposite; corolla 6-28 mm long. **(8)**

    8a. Leaves either brownish-scurfy or long-ciliate on margin, alternate; corolla lobes longer than tube.   *Rhododendron*

    8b. Leaves not as above, opposite; corolla lobes shorter than tube.   *Kalmia*

9a. Pedicels with 2 bracts just below flowers; leaves and calyces scurfy-pubescent.   *Chamaedaphne*

9b. Pedicels without bracts subtending flowers, but often bearing bracts at base of pedicel; leaves and calyces not scurfy-pubescent. **(10)**

10a. Leaves mostly 10 mm long or less, bases bulbous-decurrent on stem.   *Phyllodoce*

10b. Leaves over 10 mm long, base not bulbous-decurrent on stem. **(11)**

11a. Flowers 4-merous; plants ascending to erect, 50-150 cm tall or more, young leaves stipitate-glandular, deciduous.   *Menziesia*

11b. Flowers 5-merous; plants decumbent to erect, mostly less than 30 cm tall (trailing stems often much longer); young leaves not stipitate-glandular, or leaves evergreen. **(12)**

12a. Plants maritime; anthers 4-awned; fruit a capsule embedded in a fleshy calyx.   *Gaultheria*

12b. Plants usually not maritime; anthers 2-awned; fruit a capsule or a berry, calyx not fleshy. **(13)**

13a. Leaves entire, margins rolled under, whitish beneath; flowers in umbels, pedicels 5-20 mm long, recurved apically (at least in flower); plants in bogs.   *Andromeda*

13b. Leaves sharp to round toothed or entire, flat, not whitish beneath; flowers in racemes, pedicels 2-6 mm long, straight or curved; plants not in bogs.   *Arctostaphylos*

Andromeda polifolia   Bog Rosemary

Dwarf subshrub to 80 cm tall from creeping rhizomes with elongate, ascending branches. Leaves persistent, alternate, leathery, 2-3 cm long, linear-lance-shaped to oblong, rolled margins, lower surface white with fine waxy powder, grayish-green above with sunken veins. Flowers few, 2-6 urn-shaped on terminal clusters, nodding on red pedicels, calyx 5-lobed, petals pink to red. Fruit almost spherical, hairless, 5-valved capsule, seeds tiny, numerous.

Most of Alaska. Bogs, fens and swamps, sub alpine meadows. Common. Flowers June-July. Contains andromedotoxin which causes low blood pressure, difficult breathing, cramps, diarrhea and vomiting.

Arctostaphylos alpina   Alpine Bearberry

Prostrate, densely branched dwarf shrub 3-10 cm tall with hairless twigs, shredded papery bark and slender branchlets. Leaves deciduous, spatula-shaped to obovate with minute teeth, old leaves of previous season persist below current season leaves. Flowers bell-shaped, petals white to greenish white fading yellowish. Juicy bright red or black berry.

Lynn Canal area, Glacier Bay and Interior Alaska. Arctic and alpine tundra or woods. Uncommon. Flowers end of May to July. Berries edible when boiled, but seedy.

Arctostaphylos uva-ursi   Kinnikinnick

Creeping evergreen shrub not over 20 cm tall, forming mats by rooting along stems, twigs slender and creeping with shredding brownish-red bark. Leaves alternate, evergreen, leathery, light green, prominently net-veined. Flowers 1-to-several in raceme at ends of twigs, nodding, corolla urn-shaped, white to pink. Red, dry, seedy berry, persist into winter.

Most of Alaska except western and northwestern portions. Dry sandy places in recently deglaciated areas, dry rocky bluffs along coast and alpine. Locally abundant. Flowers June-July; fruits August. Berries edible but mealy and rather tasteless, commonly eaten by bears. Leaves used as medicinal tea to relieve kidney and bladder problems and as tobacco substitute. For tea use one teaspoon of dried leaves per cup of boiling water, steep five minutes.

Cassiope lycopodioides   Clubmoss Mtn-heather

Delicate low creeping evergreen subshrub with erect branches, stem diameter 2 mm. Leaves opposite, tiny, pressed against stem, about 6 mm long. Flowers bell-shaped, nodding on long slender stalks, sepals rounded, transparent at edges, reddish. Erect round capsule about 3 mm long.

Southeastern Alaska (Ketchikan area) west along coast to Aleutians. Exposed rocky alpine ridges. Uncommon. Flowers July to August.

Cassiope mertensiana   White Heather

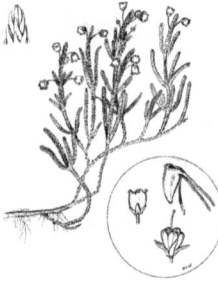

Prostrate, evergreen, mat-forming subshrub to 30 cm tall with upturned branches, 5-30 cm tall. Leaves opposite in 4 rows and pressed against stem, scale-like, grooved at base, stem diameter 4 mm, covered by leaves except at base, square in transverse section. Flowers several from sides of stem, corolla bell-shaped, 5 rounded white to pinkish sepals. Erect 5-chambered round capsule.

Southeastern Alaska except Yakutat. Mountain meadows adjacent to snowfields in alpine. Common. Flowers June-August. Branches used to produce golden dye. This species is very slow growing; individuals can be as old as 20 years so leave them be.

Cassiope stelleriana   Alpine Moss Heather

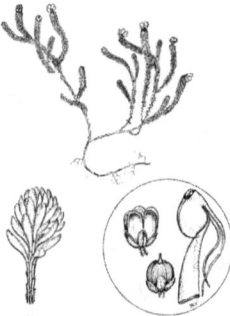

Low, spreading, mat-forming evergreen subshrub, slender, upright stems 5-15 cm tall, reddish with scattered short stiff hairs. Leaves alternate, spreading, all around stem, flat above, keeled on lower surface. Flowers terminal, mostly single, about 6 mm long, erect to nodding, corolla bell-shaped, sepals rounded, reddish, united in lower half. Flower white to pink. Erect round capsule about 3 mm long.

Southeastern Alaska west to eastern Aleutians. Protected slopes in snow deposition areas, moist seepage areas in alpine. In southeastern Alaska, less common than *C. mertensiana*. Flowers late May-July; fruits mature late July-August.

Cassiope tetragona   Lapland Cassiope

Coarse, dark-green, mat-forming subshrub 10-20 cm tall. Leaves opposite, appressed in 4 rows, lower leaf surface grooved nearly entire length, often pubescent with short white hairs, stalkless. Flowers solitary, arising from axillary buds, nodding on slender stalks, sepals whitish, bell-shaped corolla, lobes shorter than tube, white to cream. Fruit an erect round capsule.

On mainland (Juneau Ice Field) in northern end of southeastern Alaska; interior Alaska except southern coast. Alpine and arctic tundra. The most common *Cassiope* of central and northern Alaska. Flowers late May-August; fruit matures in August and September. In Interior Alaska sometimes called firemoss because even when green it burns well and used as a source of fuel where woody plants are not available.

Cladothamnus pyroliflorus   Copper-bush

Erect shrub 0.5-2 m tall with clustered long leaves, first year twigs light brown, shiny and stiff, older twigs with shredding bark. Leaves whorled, ovate to lance-shaped with rounded to abruptly pointed tip, pale green and somewhat whitish on undersides. Flowers 1- to several at ends of twigs, 5 narrow sepals; 5 salmon- or copper-colored spreading oval petals, 10 stamens, long hook near tip; long, recurved style. Round dark reddish brown many-seeded capsule.

Southeastern Alaska to Prince William Sound. Forms clumps in meadows and open forests at and just above treeline, (sometimes at low elevation) in openings, along streams, generally in association with Mountain Hemlock. Common. Flowers middle July-August; fruits August-September. One of the few heathers without bell-shaped flowers. Also known as *Elliottia pyroliflorus.*

91

Gaultheria shallon — Salal

Stiff creeping to erect evergreen shrub 0.2- 4 m tall, twigs with scattered often gland-tipped hairs, hairless with age, reddish-brown with shredding bark. Leaves stiff, leathery, large and thick, alternate, ovate to elliptic, sharply toothed on edges, short-pointed at apex, upper surface shiny green with raised veins, lower surface lighter green, short-stalked. Flowers in long glandular hairy racemes, usually at twig tips, corolla urn to bell-shaped, calyx lobes reddish brown with glandular hairs, about 1/3 as long as corolla, 10 short stamens. Flower color pink with stiff reddish brown hairs. Fruit berry-like, purplish, surrounding round capsule.

Dixon Entrance north to Sitka along coast and on islands of southeastern Alaska. Understory shrub in poorly drained western red cedar-hemlock forests, meadows and shorelines. Common in southern southeastern Alaska, in scattered locations along outer coast in northern part of range. Flowers May-June. Fruit edible and good in jam, syrups or dried in cakes.

Kalmia polifolia — Bog-laurel

Evergreen, spreading subshrub to 50 cm tall. Leaves opposite, stalkless, oblong to linear, flat or with edges rolled under, dark green above, whitish beneath, petioles short, twigs slightly two-angled. Flowers several in cymes at ends of twigs, corolla saucer-shaped with 5 lobes and 10 ridges (keels), 5 thick ovate sepals with a sparse fringe of hairs, 10 stamens. Flower color deep pinkish rose. Fruit 5-parted capsule.

Southeastern Alaska. Bogs, wet mountain meadows, often growing with bog-rosemary and Labrador tea. Common. Flowers late May-July. Toxic due to the poison andromedotoxin. Also known as *K. microphylla.*

Ledum palustre   Narrow-leaf Labrador Tea

Evergreen subshrub similar to *L. groenlandicum* but leaves smaller and narrower, rolled under at edges, twigs hairy, light brown, older twigs gray. Leaves linear to oblong usually less than 2 cm long, 0.1-0.7 cm broad, leathery, upper surface shiny, dark green, lower surface with reddish-brown woolly hairs. Flowers numerous in clusters at twig tips, 5 spreading white petals, stamens mostly 10, flower stalks with short, white hairs (rarely with some twisted, reddish hairs), sharply and abruptly bent just below apex. Long, oval, finely hairy capsule opening from base in autumn and persistent through most of winter.

Head of Lynn Canal, Glacier Bay north throughout Alaska. Bogs and black spruce forests. Common. Flowers June and early July; fruits July-August. Contains ledol, a poisonous substance causing cramps and paralysis. Old name *L. decumbens.* Now classified as *Rhododendron tomentosum*.

Ledum groenlandicum   Common Labrador Tea

Evergreen subshrub to 1.5 m tall with upright or prostrate branches, young twigs densely hairy, light brown, older twigs gray. Leaves leathery to 6 cm long, 0.3-1.5 cm broad, narrowly oblong to elliptic or lance-oblong with fragrant odor, margins strongly rolled under, underside covered with reddish-brown woolly hairs, upper surface dark green and roughened. Flowers numerous, conspicuous, fragrant in clusters at end of twigs, 5 white petals, small, protruding stamens mostly 8, flower stalks long with tangled reddish hairs, evenly curved. Hairy oblong capsule, opening from base in fall, persists through winter.

Southeastern Alaska except Yakutat area and Glacier Bay. Bogs. Common. Flowers middle June-July. Dried leaves used for tea, syrup made from tea used for coughs and hoarseness; poison at excessive doses. Now classified as *Rhododendron groenlandicum.*

Loiseleuria procumbens   Alpine-azalea

Matted or trailing evergreen subshrub 2-10 cm tall, twigs much branched, almost totally concealed by persistent leaves. Leaves opposite, elliptic, small, leathery, with margins rolled under, upper side hairless, lower side with dense short white hairs and a prominent ridge. Flowers 1 to several at twig ends, erect on stalks, corolla bell-shaped, divided into 5 lobes nearly to middle, 5 stamens, petals pink sometimes white. Erect, round, 2-3 parted dark red capsule.

Southern Alaska. Well drained rocky sites in alpine and arctic tundra; frequently forming pure mats, low elevation in bogs. Occasional to common. Flowers late May-July; fruits July-August.

Menziesia ferruginea   Rusty Menziesia

Loose-spreading deciduous shrub 2-3 m tall with slender, widely forking paired branches, twigs glandular with odor when crushed, older twigs reddish brown to gray, smooth to peeling in thin layers. Leaves oblong to elliptic, short-pointed usually with abrupt whitish tip, edges minutely toothed with gland-tipped hairs, upper side gray green with scattered brown hairs, under side whitish with glandular hairs, petioles with gland-tipped hairs. Flowers several to many at ends of twigs on glandular stalks, corolla urn-shaped with 4 shallow lobes with long glandular hairs, 8 stamens, stigma 4-lobed, petals orange to pinkish orange. Fruit 4-parted capsule, long, green to reddish brown, maturing in fall and often persists through winter.

Southeastern Alaska west to Alaska Peninsula. Conifer understory, also in openings, bogs and clearcuts. Common. Flowers late May-July; fruits September-November. Twigs woven into mats; leaves usually have thickened, fleshy, edible gall, caused by fungus (*Exobasidium*). Twigs and leaves used to make a beverage but, like bog-laurel, contains the poison andromedotoxin. This species much less palatable for deer than blueberry species; one of few forest shrubs having wind-dispersed seeds.

Phyllodoce glanduliflora Yellow Mountain-heather

Low, much branched, evergreen subshrub forming pure mats 10-40 cm tall, stems much branched, slender, with conspicuous peg-like leaf-scars. Leaves needle-like, thick with minute glandular teeth on edge, yellow-green, grooved, hairy on lower surface, crowded on upper 5-10 cm of stem. Flowers 5-15 at tips of erect or nodding stems, glandular hairy stalks, corolla urn-shaped with 5 small lobes with glandular hairs, petals yellowish or greenish yellow. Oval capsule splitting into 5 parts.

Southeastern Alaska west to Aleutians. Protected depressions adjacent to snowfields, above and below timberline, moraines and outwash at sea level. Common. Flowers early June-late August. Hybridizes with *P. empetriformis*. Also known as *P. aleutica*.

Phyllodoce empetriformis (Pink Mountain-heather) Matted much-branched subshrub 10-40 cm tall. Pink to rose colored bell-shaped flowers, sepals reddish. Alpine slopes. Juneau Icefield, head of Lynn Canal and Glacier Bay.

Rhododendron camtschaticum
Kamchatka Rhododendron

Evergreen subshrub 5-15 cm tall, twigs coarse, much-branched, gray-brown to reddish with shredding bark. Leaves obovate, tapering to base, with conspicuous stiff hairs on margins and prominent network of veins on underside, non-glandular hairs on margins, petiole lacking. Flowers 1-to -several on erect leafy stalks at ends of twigs, spreading, style conspicuous, curved, petals rose-purple to deep red or white. Capsule on a long stalk.

Scattered populations in northern southeastern Alaska including mountains above Juneau, west to Aleutians north to Denali Park and Nome. Rocky well drained sites in alpine meadows. Common in Cordova area and Aleutians. Flowers July-August. Typical subspecies (spp. *camtschaticum*) corolla hairy on outside and on margins of lobes, leaf margins mostly with non-glandular hairs. Other subspecies (spp. *glandulosum*) has a corolla without hairs on outside, margins and leaf margins with glandular hairs.

KEY TO SPECIES OF *VACCINIUM*:

1a. Leaves evergreen, margins slightly to strongly rolled under. **(2)**

1b. Leaves deciduous (at least on mature plants), flat, glandular or hairless or minutely hairy beneath; fruit blue, blue black, black, or red. **(4)**

  2a. Plants small, prostrate, evergreen, small flat rounded leaves with tiny teeth on margin, flowers never present, usually growing under dense forest canopy. *Vaccinium* immature form. Probably an immature form of dominate *Vaccinium* species such as *V. ovalifolium*, *V. alaskaense*, or *V. parvifolium*.

  2b. Plants larger with erect stems, or if decumbent then leaves shiny with entire leaf margins, often with margins rolled under, leaves ovate, obovate, or lance-shaped; plants generally occurring in bogs or other open but poorly drained sites. **(3)**

  3a. Petals reflexed, leaves 2-8 mm long, 1-4 mm broad, lance-shaped to ovate or elliptic. *V. oxycoccos*

  3b. Petals bell-shaped, not reflexed, leaves 0.4-1.8 cm long, 0.2-0.9 cm broad, oblanceolate to obovate or elliptic. *V. vitis-idaea*

4a. Leaf margins with minute teeth throughout; leaf blades oblanceolate to elliptic, leathery; plants usually less than 30 cm tall. *V. caespitosum*

4b. Leaf margins entire or partially minutely toothed, not minutely toothed throughout; leaf blades variously shaped; plants usually more than 30 cm tall. **(5)**

5a. Branchlets round in cross section; flowers 1-few from buds of previous season; calyx deeply lobed, lobes persistent on fruit; plants widely distributed in Alaska. *V. uliginosum*

5b. Branchlets slightly to conspicuously angled; flowers solitary in leaf axils of current season; calyx shallowly lobed, lobes often deciduous in fruit; plants mostly of southern Alaska. **(6)**

6a. Branchlets green, sharply angled; fruit red, calyx lobes persistent; plants of southeastern Alaska. *V. parvifolium*

6b. Branchlets brownish or greenish brown, angled but seldom sharply so; fruit blue to blue black or black, calyx lobes mostly deciduous; plants more widely distributed in southern Alaska. **(7)**

7a. Flowers appearing before leaves; leaves hairless, lacking glands on lower veins; corolla pinkish, urn shaped, longer than broad, style included; fruiting pedicels 2-8 mm long, often curved, teeth mostly on lower 1/3 of leaf; plants from southeastern Alaska west through Aleutians. *V. ovalifolium*

7b. Flowers appearing with leaves; leaves hairless or pubescent, often glandular on lower veins; fruiting pedicels 6-14 mm long, straight or somewhat curved; corolla coppery pink to white; depressed urn shaped; often broader than long, leaf margins toothed throughout; plants from southeastern Alaska west through Prince William Sound. *V. alaskaense*

Vaccinium alaskaense — Alaska Blueberry

Spreading to erect shrub to 2 m high, twigs thin, weakly angled, yellow green, becoming gray with age, ending in narrow stub. Leaves thin, oval to egg-shaped, entire, or shallowly toothed on edges, upper surface green, lower surface whitish with few short glandular hairs on mid vein. Flowers single at base of leaves after leaves partially develop, on straight stalks, corolla rounded, urn-shaped, widest just above base, usually broader than long, petals bronzy pink to white. Berry bluish black to purple, usually without bloom, on stalk straight or nearly so, enlarged just below fruit.

Southeastern Alaska west to Prince William Sound and eastern Kenai Peninsula. Coastal forests, open areas, clearcuts, usually at low elevation and associated with Early Blueberry; forms dense thickets. Very common. Flowers April-May; fruits middle July-late August. Berry edible and juicy, good flavor but not as sweet as *V. ovalifolium* which it closely resembles. Dominant forest understory shrub in coastal Alaska.

Vaccinium caespitosum — Bog Blueberry

Low spreading mat-forming shrub to 30 cm tall, twigs much branched, often rooting at nodes, young twigs slender, green with short hairs, round or sometimes angled, older twigs brown to gray, bark usually shredding. Leaves oblong to lance-shaped, rounded to short-pointed at apex, edges with fine usually gland-tipped teeth, netted veins conspicuous on some leaves, obscure on others, upper surface green, lower lighter, both hairless or with scattered stiff hairs. Flowers single at base of leaves, nodding on stalks, corolla urn-shaped with 5 small rolled lobes, petals white or pink. Sweet blue berry with bluish bloom.

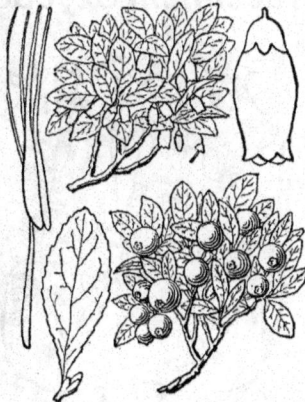

Southern Alaska. Wet meadows, bogs, poorly drained forest sites, alpine tundra, low to high elevation. Common. Flowers late May-middle July; fruits August. Berries sweet, edible and preferred over other Alaskan blueberries.

Vaccinium ovalifolium   Early Blueberry

Spreading shrub to 2 m tall, twigs slender, yellowish green to reddish, shiny, strongly angled and grooved, becoming gray by 2nd and 3rd year, ending in narrow stub. Leaves oval, rounded at tip and base, thin, entire to shallowly toothed on edges, hairless, green on upper surface, whitish beneath, leaves at tips of stems usually largest. Flowers single on sides of twig, nodding on stalks, corolla urn-shaped, broadest below middle and usually longer than broad, petals pink. Blue to bluish-black berry often with whitish bloom.

Southeastern Alaska west to Aleutians, interior Alaska north to Talkeetna. Coastal forests to subalpine, also edges of bogs, openings, clearcuts and dense thickets. Abundant. Flowers April-May before leaves emerge; fruits July-September. Berry edible with good flavor; important deer food. Sometimes impossible to distinguish from *V. alaskaense* (with which it hybridizes), but has glands along leaf midribs and darker berries.

Vaccinium oxycoccos   Bog Cranberry

Tiny evergreen shrub 10-40 cm tall with very slender stems creeping vinelike through moss and rooting at nodes, stems yellow to reddish brown, hairless when young. Leaves persistent, small, lance-shaped, short-pointed, leathery, edges rolled under, shiny dark green on upper surface, gray or whitish beneath with conspicuous midrib. Flowers 1-4 at ends of stems, nodding on erect slender stalks with two tiny bractlets below middle, 4 red to pink reflexed petals, 8 long yellow forward pointing stamens. Red juicy berry.

Most of Alaska south of 68th parallel, rarely further north. Bogs and alpine meadows, low to high elevation. Common. Flowers June-July; fruits August. Fruit edible, makes tasty jam, pies and beverages. Also known as *Oxycoccus microcarpus* and *O. oxycoccus*.

Vaccinium parvifolium Red Huckleberry

Erect shrub to 4 m tall, twigs slender, shiny green, strongly angled or ridged, ending in narrow stub. Leaves thin, deciduous, evergreen when small and stressed, often persisting on twigs into early winter, oval to elliptic, entire, green on upper surface and grayish beneath, petioles short. Flowers, single at base of leaves, nodding, corolla broadly urn-shaped with 5 small waxy lobes, yellowish pink to red. Bright red round berry.

Southeastern Alaska north including Yakutat but not found in Haines and Skagway area. Low elevation Sitka Spruce-Western Hemlock forests especially along beach terraces and on small islands, open areas, clearcuts and roadsides. Occasional to common. Flowers May-June; fruits middle to late August. Berry edible but a little tart; makes good jelly and pies, used as fish bait. Stems and bark brewed as tea and used as gargle for sore throat and inflamed gums.

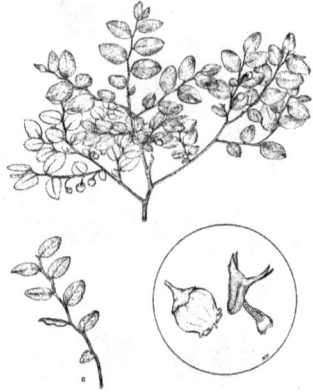

Vaccinium uliginosum Dwarf Alpine Blueberry

Much branched low shrub to 30 cm tall, erect or prostrate, often rooting along branches, twigs slender, yellowish-brown, round, minutely hairy, older twigs much branched, yellow brown to gray with shredding bark. Leaves oval to elliptic, dark green on upper surface sometimes with whitish bloom, lighter below with conspicuous veins. Flowers 1-4 from ends or side branches, nodding on stalk, urn-shaped with 4 short lobes, petals pink. Blue or black berry, often with whitish yeast bloom.

Most of Alaska. Alpine tundra to 1700 m in elevation and bogs. Common. Flowers late May-July; fruits late July and August. Berry edible, sweet, excellent in jam and pie, or used to season meat, thicken soup or add to pemmican; fresh or dried leaves good for tea.

Vaccinium vitis-idaea   Lingonberry

Evergreen, creeping, mat-forming subshrub to 15 cm tall, stems slender and trailing, rooting at nodes, light brown to yellow. Leaves about 1 cm long, oval, wide, thick, green and shiny above, light green beneath with black dots and short stiff brown hairs, edges slightly rolled under. Flowers 1 to several, nodding on short stalks at twig ends, corolla bell-shaped with 4 short lobes, pink. Bright red sour berry.

Most of Alaska. Bogs, alpine tundra, poorly drained forests, loose mats in moist mossy sites, dense mats on dry rocky slopes in alpine, low to high elevation. Common. Flowers middle June-July; fruits August. Berry edible and excellent in jam and makes a tasty beverage; chewed to cure headache, sore throat, upset stomach; boiled and rubbed on rashes including measles.

FABACEAE
PEA FAMILY

KEY TO GENERA OF *FABACEAE:*

1a. Leaves even-pinnate. **(2)**

1b. Leaves odd-pinnate, palmately compound, or simple. **(3)**

    2a. Styles bearded down one side; plants generally coarse.   *Lathyrus*

    2b. Styles bearded at apex; plants generally slender.   *Vicia*

3a. Leaflets toothed (almost entire in some *Trifolium*); leaves palmately trifoliolate.

    Trifolium

3b. Leaflets (or leaves) entire; leaves pinnately or palmately 5-many-foliolate, or simple. **(4)**

4a. Leaves palmately compound; stamens united by their filaments into a tube or column.   *Lupinus*

4b. Leaves pinnately compound or simple; stamens formed in two groups through union of their filaments. **(5)**

5a. Fruit a loment (a flat legume constricted between seeds, falling apart at constrictions when mature into one-seeded joints); wings shorter than keel.

    Hedysarum

5b. Fruit a legume (a fruit formed from one carpel that splits from two lines of opening); wings mostly as long as or longer than keel. **(6)**

6a. Keel with an apical beak directed outward and forward; ventral suture of pods produced internally forming a partial or complete partition, or partition absent; plants mostly stemless (with a stem in some *O. deflexa*).   *Oxytropis*

6b. Keel lacking a beak; sutures variously produced internally, or not produced at all; plants mostly with well developed stem.   *Astragalus*

Astragalus alpinus   Purple Alpine Milk-vetch

Perennial herb to 40 cm long, matted, stems creeping at base, stipules 3-10 mm long, ovate to triangular, lower ones united-clasping, upper ones nearly free. Leaves ovate to elliptic or oblong with short, stiff, appressed hairs below, 13-23 bright yellow-green leaflets per leaf. Flowers clustered atop long stem in elongated racemes with 2-many flowers, erect at first, later spreading and finally reflexed in age, petals violet to dark pink purple, often fading yellow on drying, calyx black-hairy. Seedpod with black hairs, dorsal suture grooved.

Most of Alaska except Aleutians. Alpine, open areas, grassy slopes, gravel moraines and scree slopes. Common. Flowers late June-August. A variable and widely distributed species.

Astragalus eucosmus (Elegant Milk-vetch) Perennial herb 25-55 cm tall from stout woody rhizome, stems glabrous or nearly so, stipules lance-shaped to oval, leaflets 13 or 15, narrowly oblong to elliptic; flowering raceme first dense then elongated, calyx black-haired with linear-lance-shaped teeth, corolla purple, wings slightly longer than keel; pods stalkless or short stalked, reflexed, densely black hairy, rarely white-hairy. Sandy, gravelly, or clay soil in woods or thickets, less common in tundra, found in northern southeastern Alaska and from Seward Peninsula northward and eastward through most of interior Alaska. Flowers June-July. This and other ***Astragalus*** species may be toxic. They absorb large quantities of selenium and molybdenum.

Astragalus robbinsii (Robbins Milk-vetch) Perennial to 50 cm tall, stout stem, stipules lance-shaped, leaves with 9-15 oblong leaflets; racemes to 20-flowered, calyx with black, stiff, closely pressed, short hairs, petals yellow-whitish (sometimes fading bluish) with tip of keel purple, keel shorter than wings, much shorter than banner; pod ellipsoid. Terraces and floodplains in woods and thickets disjunct from Juneau area to Glacier Bay, Denali Park and along Yukon River. Flowers June-July. Includes species ***A. harringtonii***.

Hedysarum alpinum   Alpine Sweet-vetch

Perennial 20-70 cm tall from a woody base and taproot, stems with stiff, short, simple, flat lying hairs, stipules brownish and united. Leaflets 9-21, lance-shaped to oblong. Racemes with several to many flowers, keel longer than wings or banner, calyx bell-shaped. Loments 5, constricted between seeds into segments that split transversely.

Lynn Canal to Glacier Bay north throughout Alaska. Glacial moraines, river bars, rocky slopes, arctic and alpine tundra and woods. Flowers June-July. Root eaten raw, boiled or roasted; also bear and moose food. *H. mackenzii* an interior species found near Haines reported but not proved poisonous. This is the infamous supposedly poisonous plant eaten by Chris McCandless of *Into the Wild*.

Lathyrus japonicus   Beach-pea

Perennial herb to 1.5 m tall, hairless or sparsely hairy from slender rhizome; angled, prostrate stem. Leaves tipped with curling tendrils, thick, fleshy, with 6-12 oblong to ovate leaflets, stipules resemble leaves. Raceme few-flowered, hairless or sparsely hairy, calyx and peduncle reddish brown, reddish banner and bluish-violet wings and keel. Pods usually pubescent.

Along coast and on islands in much of Alaska; rare inland. Sandy beaches. Common. Flowers May-August. Also known as *L. maritimus*. The seed contains a toxic amino-acid which, in large quantities, can cause a serious disease of the nervous system known as "lathyrism". If eaten, eat in small quantities.

Lathyrus palustris   Wild-pea

Perennial herb to 1 m tall from slender rhizome, plant hairy, stem winged. Leaves 4-8, lance-oblong to linear, usually with 3 pair of linear leaflets, tendrils branched and clasping, wingless petioles. Flowers 2-8, upper calyx teeth broadly triangular, shorter than lance-shaped lower teeth, petals pink to pink-purple, rarely white. Pods 3-6 cm long, hairy to smooth.

Southeastern Alaska west to Aleutians, in widely scattered sites in interior Alaska to 66th parallel. Wet meadows, beaches and tidal flats. Most common in southern and southeastern Alaska. Flowers June-July.

Lathyrus ochroleucus (Cream-flowered Pea) Cream to yellowish flowers and a hairless fruit. Woods and thickets only from Hyder in southeastern Alaska. Flowers May-July.

Lupinus nootkatensis Nootka Lupine

Perennial herb to 60 cm tall, stems from stout, hollow, branched, woody base, long woody root. Leaves alternate, each with 5-8 leaflets radiating from a common center, obtuse to rounded or blunt-tipped, silky and hairy. Flowers in dense clusters at top of plant, as many as 5 petals, racemes 5-35 cm long, lower broad, boat-shaped lip 6-12 mm long, flower color blue often shaded pink or white. Fruit black, hairy or silky pods, 7-11 seeds.

Southeastern Alaska west to Aleutians, extends into Interior as far as Cantwell. Open areas, beaches, gravel bars, from low elevation to alpine. Common. Flowers July-August. Seeds poisonous causing inflammation of stomach and intestines. Rhizome can be fatal if eaten raw. *L. lepidus* (Prairie Lupine) silky-hairy and less than 40 cm tall with closely flowered racemes found at Hyder.

Lupinus arcticus (Arctic Lupine) Perennial herb, to 60 cm tall, pubescence closely pressed to spreading, leaflets narrower and acute at tip; racemes shorter (4-14 cm long), lower slender lip 6-11 mm long, flowers blue; 5-8 seeds. Arctic and alpine tundra and woods from Juneau to Lynn Canal north throughout continental Alaska; also reported on Pribilof Islands. Flowers June-July. Probably poisonous; hybrids with *L. nootkatensis* occur. Also known as *L. latifolius*.

Oxytropis campestris Field Locoweed

Tufted perennial herb covered with silvery hairs. Leaflets lance-elliptic to oblong, opposite to whorled stipules lance-shaped. Bottlebrush spike of 8-12 yellowish-white, yellowish or purple-tinged flowers on leafless stems, flowers well above leaves. Pod erect, stalkless with soft spreading hairs.

Southeastern Alaska west to Aleutians and north to 66th parallel. Dry sandy places, alpine; sand dunes in Yakutat. Uncommon. Flowers June-July. Recently split into several species.

Oxytropis deflexa (Deflexed Oxytrope) Short-stemmed, prostrate to erect, 15-40 cm tall. Leaflets 23-45, crowded, ovate to narrowly lance-shaped or oblong, sparsely to densely soft-haired, stipules united; flowers dingy white, pinkish, bluish, or bright pink purple, calyx bell-shaped; pods drooping, straight, curved, or oblong. Tundra in much of Alaska except western and southwestern parts. Flowers June-July.

Oxytropis nigrescens (Blackish Oxytrope) Pubescent with long, gray hairs, tufted, woody base covered with persistent stipules and petioles, 7-13 oblong to oval leaflets, pubescent on both sides and margins fringed with hairs, stipules with lance-shaped free parts, margins fringed with hairs; inflorescence 2-3- flowered, calyx densely black-haired with linear teeth about as long as tube, petals purplish to blue or white; pods almost stalkless, oblong, gray to black pubescent. Arctic and alpine tundra in most of Alaska. Similar species *O. huddelsonii* 5-19-foliate, leaflets with more or less rolled margins, white-hairy above, sparsely hairy below, stipules whitish, pods elliptic with hooked beak. This species much overlooked.

Oxytropis scammaniana (Scamman Oxytrope) Densely tufted with persistent straw-colored stipules, 7-19 lance-shaped to elliptic leaflets, covered with short, soft, spreading hairs on both sides or hairless above, stipules broad with elliptic, obtuse, free lobes; flower scapes with white, soft, spreading hairs, racemes few-flowered, calyx with soft, black, spreading hairs and linear, acute teeth, petals bluish or purplish; pods short, thick, erect, black-haired. Arctic and alpine tundra, rarely in southeastern Alaska; in interior Alaska from Brooks Range and Alaska Range east to southern Yukon. Flowers June-July.

Oxytropis viscida (Viscid Oxytrope) Caespitose, tufted, 19-51 lance-oblong leaflets often glandular and warty, stipules pale, more or less glandular-warty, margins fringed with hairs, bracts linear-lance-shaped; many-flowered racemes formed in heads or elongate, calyx teeth glandular; pods black-pubescent. Sandy and gravelly soils, in tundra, woods and on rock outcrops in most of Alaska. Flowers June-August. Highly variable species. Also known as *O. borealis.*

104

Trifolium wormskjoldii   Coast Clover

    Perennial prostrate herb from taproots, often rhizomatous, hairless, stipules sharply toothed, stems to 30 cm long. Petioles longer than oblong, blunt-tipped leaflets. Flat spine-pointed bract below flower heads, calyx lobes needle-like, petals pale red-purple to white. Pods 1-4 seeded.

    Southern southeastern Alaska. Meadows and coastal dunes. Uncommon. Flowers June-August. Rhizomes dug in early spring or after frost in fall and dried, roasted, or boiled and eaten; sweet tasting like green peas.

Vicia nigricans   Giant Vetch

    Climbing vigorous perennial herb, stems to 2 m tall, somewhat swollen, stipules toothed. Leaflets 14-32, oblong to lance-shaped. Flowers on peduncles in racemes, flower color yellowish tinged with red or purple (or orange). Pods 3-5 cm long.

    Coastal southeastern Alaska and Cook Inlet. Beaches, meadows and along streams. Occasional. Flowers June-July. Some indigenous people ate seeds but they are considered poisonous. Also known as *V. gigantea*.

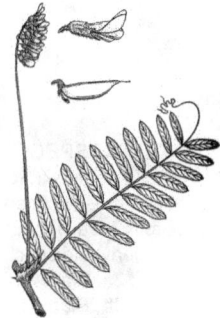

    Vicia americana (American Vetch) Perennial; stems to 1 m tall sparsely hairy with 8-16 nearly elliptic leaflets tipped with a sharp firm point and a few serrations, 15-45 mm long. Flowers in short racemes 2-8 flowered, pink to pink purple. Fruit a glabrous pod 3-4 cm long. Open woods, thickets and meadows mostly in southeastern Alaska. Flowers June-July.

FUMARIACEAE
FUMITORY FAMILY

Corydalis aurea   Golden Corydalis

Annual or biennial herb with many branched stems 10-60 cm long covered with whitish bloom. Leaves 1-4 times pinnately compound. Racemes several-flowered, pedicels longer or almost equal to spur, sepals yellowish or whitish, flower color golden yellow. Capsules linear with long style, usually curved and constricted between smooth seeds.

Lynn Canal area in southeastern Alaska, eastern Interior south of 66th parallel. Roadsides and sandy places. Occasional. Flowers June-July.

GENTIANACEAE
GENTIAN FAMILY

KEY TO SPECIES OF *GENTIANA:*
1a. Corolla tube with folds between lobes; nectary glands at base of ovary; stem leaves mostly conspicuously connate-sheathing. **(2)**
1b. Corolla tube without folds between lobes; nectary glands at base of corolla tube; stem leaves distinct, or lower ones slightly connate. **(5)**
   2a. Plants annual (or biennial), from slender taproots. **(3)**
   2b. Plants perennial, from rhizomes or woody bases. **(4)**

   3a. Plants prostrate to ascending, simple or branched from base; flowers solitary, terminal.   **G. prostrata**
   3b. Plants erect or ascending, branched throughout or simple; flowers terminal and axillary, often more than 1 and often white.
       G. douglasiana
   4a. Calyx 2-lipped, one lip 2-toothed, other 3-toothed; corolla usually bright blue, lobes orbicular-acuminate; stems usually with more than 3 pair of leaves below bracts.   **G. platypetala**
   4b. Calyx almost equally 5-toothed; corolla cream to yellowish green or blue, rarely white, lobes merely acuminate or ovate; stems usually with 3 pair of leaves below bracts or fewer.   **G. glauca**
5a. Sepals nearly distinct; pedicels longer than next internode; each corolla lobe bearing two fringed scales.   **G. tenella**
5b. Sepals connate for one-fourth to one-third their length; pedicels shorter than next lower internode (except on some lateral branches); each corolla lobe with a continuous fringe, or naked at base within.   **G. amarella**

Gentiana amarella — Northern Gentian

Annual herb 10-50 cm tall, stem single or branched, erect or ascending from a taproot, 3-6 internodes. Leaves 5-8 pairs, lance-oblong to narrowly egg-shaped. Flowers in several to many cymes or axils, pedicels shorter than next lower internode, corolla tubular, 5-lobed, lobes longer than tube, flower color violet or lilac, rarely white; flowers do not have folds between petal lobes, lobes basally frilly. Capsule cylindrical, seeds yellow and smooth.

Southern Alaska. Moist places, meadows, and along streams. Occasional. Flowers July-August. This species also called *Gentianella amarella* and forms hybrids with *Gentianella propinqua*. Found on Mt. Roberts near Juneau.

Gentiana douglasiana — Swamp Gentian

Hairless taprooted herb 5-25 cm tall, stem angled, mostly branched. Egg-shaped basal leaves forming small rosette, stem leaves egg-shaped to elliptic, smaller than basal leaves. Flowers solitary or in cymes, closely subtended by ovate bracts and often in axils of leaves as well, calyx 5-toothed, lobes shorter than tube, corolla tubular bell-shaped. Tube yellowish green, lobes whitish, spotted and streaked with purple on outer petal, whitish folds. Capsule oblong, flattened, stalkless, numerous black seeds.

Southeastern Alaska west to Kenai Peninsula. Bogs, fens, meadows low elevation to alpine. Common in southeastern Alaska. Flowers July-September. Named after famous Scottish botanist and explorer David Douglas.

Gentiana glauca — Glaucous Gentian

Perennial herb 4-15 cm tall, erect, simple, stem hairless, yellow-green. Rosette leaves obovate, somewhat fleshy, stem leaves in 1-3 stalkless pairs, elliptic to rounded. Flowers in terminal cymes, enveloped by upper leaves, calyx 5-toothed, lobes shorter than tube, corolla tubular, blue to yellowish-green, folds pale. Capsule lance-ovoid.

Northern southeastern Alaska to Prince William Sound. Arctic and alpine tundra. Occasional. Flowers July-August.

107

Gentiana platypetala   Spotted Gentian

Rhizomatous hairless perennial herb 10-40 cm tall from an unbranched, thick, horizontal rhizome, rosette leaves lacking. Stem leaves only, in several opposite pairs, egg-shaped to elliptic, becoming larger above. Flowers solitary, stalkless, calyx 2-lipped, one lip 2-toothed, other 3-toothed, corolla funnel-like, bright blue, spotted with green inside. Fruit elliptic-oblong capsule, small numerous seeds.

Southeastern Alaska west to Kenai Peninsula and Kodiak Island. Wet meadows from sea level to alpine. Common. Flowers July-September.

Gentiana tenella   Slender Gentian

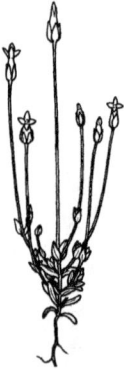

Slender annual herb 2-10 cm tall with 4-angled stem. Basal leaves elliptic to obovate; stem leaves lance-shaped to elliptic. Flowers solitary at end of each branch or in axils of leaves, 4(5) petals, lobes fringed within, blue or white. Capsule narrow, slightly exceeding corolla; seeds yellow, egg-shaped, nearly smooth.

Juneau area and from islands of Bering Sea, disjunct eastwad through eastern Alaska and north along coast. Arctic and alpine tundra. Occasional. Flowers July-September. Also known as *Gentianella tenella*.

Gentiana prostrata (Pigmy Gentian) Annual to 8 cm tall. Stems low, usually more or less growing along the ground without rooting. Leaves numerous, generally oval-shaped and up to about six millimeters long, faintly white-margined, mainly in pairs on stem. The solitary, tubular, terminal flower is about a centimeter wide at the mouth, with triangular or diamond-shaped lobes in shades of deep blue to purple. Fruit a capsule containing wingless seeds. Much of Alaska. Wet mountain meadows. Rare. Flowers July-August.

Lomatogonium rotatum   Marsh Felwort

Annual herb 5-50 cm tall, hairless, stem without basal rosette. Leaves opposite, basal leaves elliptic to flattened spoon-shaped, withering early; stem leaves ovate to linear-lance-shaped. Flowers showy, blue with darker veins, white with blue strips or merely white, borne in terminal cymes or leaf axils, calyx lobes linear, as long as petals, acute, corolla wheel-shaped, (4)5 lobed, each with 2 scaly basal appendages. Capsule oblong, somewhat acute with many hairless seeds.

Mostly northern southeastern Alaska west to Aleutians. Bogs, marshes, along streams and fresh water shores mid to high elevation. Occasional. Flowers July-August. This is an extremely variable species.

Swertia perennis   Swertia

Rhizomatous perennial herb 10-60 cm tall, stem woody at base. Basal leaves egg-shaped to oblong-elliptic, long-stalked; stem leaves similar, short-stalked to stalkless, reduced upwards. Flowers dark bluish purple, often spotted and striped, in terminal and axillary cymes, 5 wheel-shaped petals. Capsule narrowly egg-shaped; seeds flattened, dark.

Coast and islands of southern Alaska. Bogs, meadows, wet open areas and subalpine meadows, low to high elevation. Occasional. Flowers July-August.

GERANIACEAE
GERANIUM FAMILY

Geranium erianthum   Northern Geranium

Perennial herb 30-80 cm tall, stems from a long, thick, light-brown scaly rhizome. Basal leaves with closely pressed hairs, long-petiolated, leaf blades flat with circular outline, much broader than long, segments not distinct. Sepals hairy, bristle-tipped, 5 large hairless petals, blue to pink purple, rarely white. Fruiting pedicels mostly once to twice longer than calyx, beak of style 2.5-5 mm long in fruit, hairless or with stiff, short, flat lying hairs at base.

Southeastern Alaska west to Aleutians north to 65th parallel. Stream sides, meadows, rocky areas, beaches and alpine. Common. Flowers June-August. Leaves boiled as astringent tea. Roots used in medicinal tea for sore throat, ulcers, diarrhea and heart trouble. For mouth sores chew raw root.

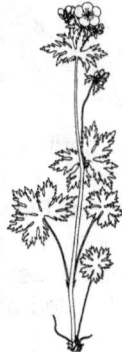

GROSSULARIACEAE
CURRANT FAMILY

Ribes bracteosum   Stink Currant

More or less erect shrub to 3.5 m tall, unarmed, stems and leaves with yellow glands, strong skunky odor. Leaves large, 5-7 lobed, toothed at edges, dotted with tiny resin glands below. Flowers long, erect to ascending racemes 15-50-flowered with leaf-like bract at base, sepals white to greenish-white, sometimes strongly tinged purplish-brown. Berries blue-black and glandular with whitish bloom.

Southeastern Alaska west to Prince William Sound. Stream banks, forest edge and wet gravelly areas. Occasional. Flowers May and June, fruit ripens late July and August; taste from unpleasant to excellent. Mixed with salmon roe and stored for winter use. Hybrids of *R. bracteosum* and *R. laxiflorum* known from Juneau area. Distinguished from other *Ribes* species by its long clusters of flowers and fruit.

Ribes lacustre   Bristly Black Gooseberry

Usually a spreading or sometimes erect shrub to 2 m tall with spiny twigs, twigs yellowish brown, densely to sparsely covered with numerous small, sharp golden prickles with larger thick spines at leaf nodes. Leaves deeply dissected, maple-like, petioles with bristly hairs. Flowers 6-15, reddish to maroon on drooping raceme, sepals oval, covered with gland-tipped hairs, light green to purplish. Berry black to dark purple, bristly with gland-tipped hairs.

Along coast and on islands of southeastern and south-central Alaska, also central and eastern Interior Alaska. Forests, along streams and on beaches. Occasional in isolated clumps. Flowers June-July; fruits August. Berry edible, but due to its odor, low production and bristleness, not much used. Spines may cause a serious allergic reaction in some individuals. Bark peeled and boiled as tea to drink during labor or as a wash for sore eyes.

Ribes laxiflorum   Trailing Black Currant

Decumbent to erect pungent shrub to 1 m tall, often rooting along stem, unarmed, bark reddish brown. Leaves maple-like, divided into 5 deeply triangular lobes, deeply heart-shaped at base. Flowers 6-18 in long, greenish-white to reddish-purple erect clusters, shorter than leaves; pedicels, rachis, and hypanthium covered with red stipitate-glandular hairs, sepals petal-like, broadly ovate, petals greatly reduced (about 1 mm long), wedge-shaped and opposite sepals, sepals and petals reddish to purple. Berries purplish black, glandular-bristly, strong odor when crushed.

Southeastern Alaska west to Kenai Peninsula. Open forest, forest edge, streams, wet areas, roadsides, avalanche tracts and clearcuts. Common. Flowers early to late May; fruits late July to early August. Berry edible but not choice. Bark, leaves, twigs and root used as a medicinal tea for colds.

Ribes glandulosum (Skunk Currant) Similar to *R. laxiflorum* in habit and leaf characteristics. Unarmed sprawling or reclining shrub; leaves 5-7-lobed; racemes ascending, 7-10-flowered, petals white or pink (red); berries dark red with a fetid odor and flavor. From tundra to woodlands, mostly in interior southern Alaska and less commonly in coastal southwestern Alaska but also Yakobi Island in northern southeastern Alaska.

Ribes hudsonianum (Northern Black Currant) 50-150 cm tall, with light-gray twigs, unarmed, glandular throughout with stalkless yellow crystalline glands, leaf blades 3-5-lobed; petals white, sepals whitish, mostly 3-veined; berries black, covered with a whitish bloom, somewhat glandular with yellow glands and barely edible. Occasional in woodlands and thickets throughout Alaska. Flowers in June.

HALORAGACEAE
WATERMILFOIL FAMILY

Hippuris vulgaris   Common Mare's-tail

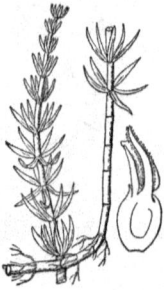

Aquatic hairless herb 10-40 cm tall, aerial stem weak and limber from stout creeping rhizome. It roots underwater, but most of its leaves are above the water surface. Leaves 6-12 in a whorl, linear, stalkless, entire, acute, those under water are thinner and limper and longer than those above water, especially in deeper streams. Flowers tiny, inconspicuous, stalkless in axils of upper leaves, petals lacking, anthers almost equal to pistil in size, not all plants produce flowers. Fruit nut-like, non-splitting, 1-seeded.

Most of Alaska. Usually at least partially submerged in fresh or brackish water. Common. Flowers July. Shoots used as potherb or for soups; plant parts tender and can be gathered at any stage of growth.

Hippuris tetraphylla   Four-leaf Marestail

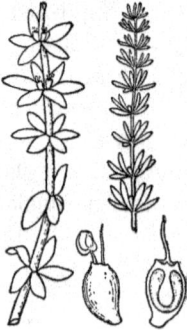

Aquatic herb, able to spread vegetatively by its horizontal rhizome; tends to be coarser than *H. vulgaris*, stem reddish at base. Leaves elliptic to oblong-egg-shaped, rounded at tip, 4-6 per whorl. Both stamens and pistils mostly functional, anthers almost equal to pistil in size. Fruit nut-like, about 2 mm long.

Southeastern Alaska west along coastal margin of Alaska. Brackish pools, tide flats and mud flats. Common. Flowers June.

Hippuris montana (Mountain Marestail) Small aquatic herb to 10 cm tall; leaves less than 10 mm long, 5-8 to a whorl; flowers mostly lacking stamens and pistils, staminate ones well below pistils, anthers stalkless or on filaments. Rare. More terrestrial than aquatic. Subalpine wet meadows, on mossy stream banks from southeastern Alaska to Aleutians. Flowers July-September.

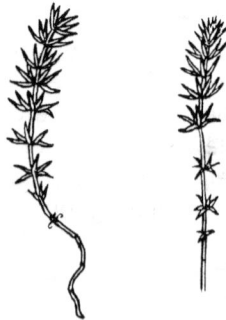

Myriophyllum spicatum   Spike Watermilfoil

Aquatic perennial immersed or emerging above water, stems 30-100 cm long, rhizomatous. Leaves whorled 3 or 4 per node with feather-like dissections. Flowers tiny, unisexual in bract axis, 4 sepals, petals 4 or falling early, stamens 4-8. Fruit 4 1-seeded achenes or nutlets.

Much of Alaska. Ponds and slow-moving streams. Common. Flowers June-August. The fruits and leaves can be an important food source for waterfowl thought to be an important source of seed dispersal. May be introduced; includes **M. sibiricum**. **M. verticillatum** with pinnately divided floral bracts is a native species.

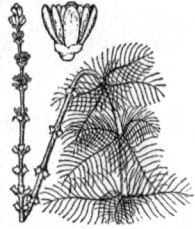

Myriophyllum alterniflorum (Alternate-flowered Watermilfoil) Fresh-water aquatic. Leaves usually 4 in a whorl, 1-2.5 cm each, pinnate with linear segments giving feathery effect. That much in common with **M. spicatum**, but flower-spike differs, upper (male) flowers in ones or twos, not whorls of 4. Lowest flowers are female in whorls of 2-4, above these are hermaphrodite flowers, and at the top male flowers in ones or twos. Petals yellow streaked with red. Flower-spike has drooping tip in bud, unlike **M. spicata**; bracts mostly opposite and shorter than flowers. Reported from widely disjunct sites in Alaska.

HYDROPHYLLACEAE
WATERLEAF FAMILY

Phacelia franklinii   Franklin's Phacelia

Perennial herb 10-60 cm tall often branched from base and sometimes above, usually purplish and hairy, densely in inflorescence. Leaves mainly on stem, slightly reduced upwards, blades once or twice pinnatifid. Flowers cymes solitary or several at ends of stems, corolla deciduous, hairless inside, hairy outside, filaments hairy, stamens included or slightly protruding, petals purplish. Capsules 6-10 mm long with 2-several seeds in each of 2 locules.

Skagway and central eastern Alaska. Disturbed areas, sandy soil. Uncommon. Flowers June-July.

Phacelia mollis (White Phacelia) An attractive plant, greenish and equally hairy in inflorescence, leaves mainly basal; corolla cream to yellowish, often tinged with blue, persisting or deciduous, hairy inside and out, stamens long-exerted; filaments hairless. Sandy or gravelly soils or rock outcrops and open woods near Haines and Skagway and central eastern Alaska.

Phacelia sericea (Silky Phacelia) Greenish, mainly basal leaves, stem leaves much reduced upwards; corolla purple or dark blue, persistent, hairy on inside and outside, stamens long and protruding, filaments hairless. Found rarely in open woods in scattered locations such as Klukwan in southeastern Alaska and eastern Alaska (Eagle). Flowers June-August. Evidently rare in our region.

Romanzoffia sitchensis Sitka Mist-maid

Short tufted perennial with stems to 30 cm long, sparsely hairy. Leaves with long stalks, basal leaves round to kidney shaped, shallowly cleft or toothed, hairless, petioles dilated at base. Flowers borne in raceme-like cymes, corolla bell-shaped, 5 united petals, sepals distinct or nearly so, flower color cream to white with a golden eye. Capsules 5-6 mm long with numerous seeds in each of two locules.

Southeastern Alaska west to Kodiak Island and northern Alaska Peninsula. Moist places from sea level to lower alpine, rock outcrops and along streams. Occasional. Flowers June-August.

Romanzoffia unalaschcensis (Alaska Mist-maid) Similar to *R. sitchensis,* to 10 cm long, but coarser and pedicels shorter or about as long as calyx, corolla cream to white, slightly larger than calyx. Moist banks and crevices of rock outcrops, ocean bluffs; southeastern Alaska west to Aleutians. This genus is similar to *Saxifraga* and can be mistaken for a member of that genus, but fruit is a round capsule rather than a follicle. Also known as *R. tracyi.*

LAMIACEAE
MINT FAMILY

Lycopus uniflorus   Northern Bugleweed

Perennial 10-50 cm tall, stem erect, square, from tuberous rootstock, thread-like runners at base. Leaves hairless, opposite, lance-shaped to lance-oblong, gradually narrowed at both ends, sharp toothed. Flowers stalkless, borne in dense, axillary clusters subtended by foliage leaves, calyx lobes triangular, hairless with age, petals white or tinged pinkish. Nutlets ridged with toothed tip.

Southern southeastern and central Alaska (Manley Hot Springs). Marshy areas, stream banks and lake shores. Uncommon. Flowers July-August.

Mentha arvensis   Field Mint

Rhizomatous, perennial, aromatic herb 15-80 cm tall with square stems. Leaves oval to lance-shaped, toothed. Flowers in axillary whorls with leaf-like whorls reduced upwards, calyx teeth triangular, corolla hairy on outside, petals lilac or white. Fruit dry, breaking at maturity into 4, 1-seeded nutlets.

Southeastern Alaska west to Alaska Peninsula north to 66th parallel. Wet areas, lake shores and riverbanks. Flowers July-August. An extremely variable species. Used to make mint tea, good for digestion.

Prunella vulgaris   Self-heal

Perennial herb with stems to 10-50 cm tall from taproots, often rhizomatous; pubescent with multicellular hairs. Leaves oval, margins with rounded teeth or entire. Flowers in thick, oblong spike of purple-brown bracts, calyx of 2 long and 3 spur-tipped lobes, petals pink purple or pink to white. Fruit dry, splits into two halves, beaked at maturity.

Southeastern Alaska, Prince William Sound, scattered in Aleutians. Meadows, disturbed areas and lake shores. Occasional. Flowers July-August. Hybrid of native and introduced species. Brewed into tea for heart problems.

Scutellaria galericulata   **Marsh Skullcap**

Perennial, rhizomatous herb 20-90 cm tall, square stems. Leaves lance-shaped to oblong egg-shaped, irregularly sharply toothed with sparse rough hairs or bristles. Calyx hairy, upper lobe distinct from lower lobe, petals pink purple to blue or white. Fruit 4 warty yellowish nutlets.

Juneau north to head of Lynn Canal, southeastern third of mainland Alaska. Margins of ponds, lakes, and streams; marshes and bogs. Uncommon. Flowers June-August.

Stachys palustris   **Swamp Hedge-nettle**

Stems 20-70 cm or more tall, erect, simple or branched, leaves tend to be stalkless. Leaf blades lance-oblong to lance-elliptic, somewhat clasping, margins with rounded to sharp forward pointing teeth with rough hairs or bristles. Flowers in interrupted spikes, calyx long-hairy with glands on stalks, lobes shorter than tube, petals purplish, white-spotted. Fruit dry, breaking into 4, 1-seeded nutlets.

Southern, southeastern and central Alaska. Stream and lake shores, meadows and other moist places. Common. Flowers June-July. Hedge-nettles give off a fishy odor when crushed.

Stachys mexicana (Mexican Hedge-nettle) Perennial 30 -100 cm tall, distinctly petiolated leaves, blades lance-egg-shaped; corolla pink purple to pink. Moist open woods, clearings. Reported from extreme southeastern Alaska (Annette and Prince of Wales Islands). Flowers June-July. Also known as *S. emersonii* and *S. ciliata.*

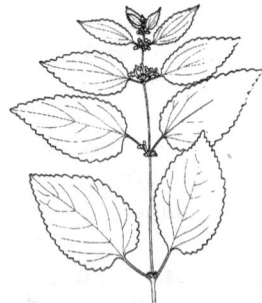

LENTIBULARIACEAE
BLADDERWORT FAMILY

Pinguicula villosa   Hairy Butterwort

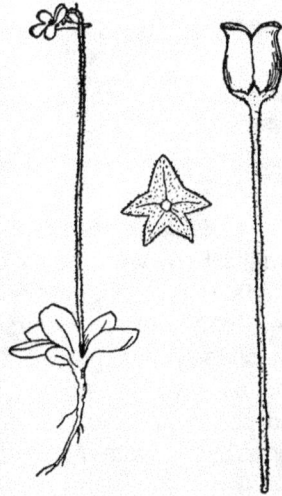

Insectivorous perennial 3-13 cm tall, scape single, long, soft hairy below, glandular above. Leaves in basal rosette, each leaf 0.4-1.5 cm long, 0.2-0.7 cm broad with sticky surface above and strongly rolled margins. Flowers nodding on softly hairy stalks, corolla of 5 united petals, lower lip 2-lobed, upper one 3-lobed and produced into a basal spur, flower color blue violet, rarely white. Fruit erect capsule ovate to spherical, 3-5 mm long.

Most of Alaska. Sphagnum bogs, fens and wet meadows. Occasional, less common than *P. vulgaris*. Flowers July. Tends to grow in nitrogen-poor soil; sticky leaves trap and digest flying insects for nitrogen. Butterworts were believed to protect humans from fairies and witches.

Pinguicula vulgaris   Common Butterwort

Insectivorous perennial, scape 3-16 cm tall from light-colored fibrous roots. Leaves in basal rosette, each leaf 1-5 cm long, 0.8-2.4 cm broad, succulent with rolled margins, broadly lance-shaped to elliptic. Flowers nodding, violet-like with white hairs in throat, spur long, blunt, scape mostly hairless, flower color blue to violet, rarely white. Fruit spherical capsule.

Most of Alaska. Sphagnum bogs, wet open areas near beaches, low to subalpine elevation. Common. Flowers late June-early August. *P. vulgaris* larger, a more or less hairless version of *P. villosa*.

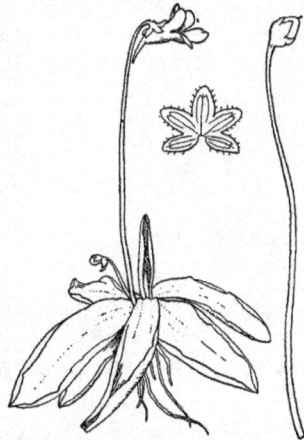

117

Utricularia vulgaris Common Bladderwort

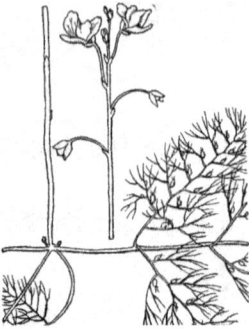

Aquatic apparently rootless perennial, stems coarse and usually floating. Leaves alternate, divided into numerous, thread-like segments with bladders. Flowers in emergent scapes, 2-5, spur about as long as lip, petals yellow. Fruit a capsule with downward bent pedicels.

Southeastern Alaska west to Aleutians north to 70th parallel, less common northward. Ponds, lakes and bogs. Common. Flowers July-August. Bladderworts are carnivorous, their bladders trap crustaceans and other aquatic organisms. Also known as *U. macrorhiza*.

Utricularia intermedia (Flat-leaf Bladderwort) Aquatic herb with stems growing along bottom in shallow water, leaves dissected into flat, linear segments, bladders on separate leafless branches; yellow flowers have a spur about as long as lower lip; fruiting pedicels erect. Lakes and ponds in much of Alaska except Aleutians. Forms hybrids with *U. minor*. Flowers July-August.

Utricularia minor (Lesser Bladderwort) Aquatic herb with stems growing along bottom of ponds or floating, bladders borne on some but not all leaves; flower spur short and sac-like, or lacking; fruit pedicels bent downward. Shallow ponds in southeastern two thirds of Alaska. Flowers July-August.

LINACEAE
FLAX FAMILY

Linum perenne Wild Flax

Perennial 20-70 cm tall arising from stout taproot. Leaves linear to narrowly oblong, 1-veined, sharp pointed. Flowers showy, axillary and terminal, 5 styles longer than or equal to stamens, petals blue. Fruit depressed globe-shaped capsules.

Juneau north to eastern half of continental Alaska. Dry open areas and roadsides. Occasional. Flowers June-August. Highly variable species.

LORANTHACEAE
MISTLETOE FAMILY

Arceuthobium campylopodum   Dwarf Mistletoe

Small hairless yellowish to greenish brown parasitic dwarf shrub on twigs, lower branches and trunks of hemlock with fragile jointed angled stems. Leaves reduced to paired brownish scales, joined at base in ring around twig. Flowers minute, paired and stalkless or nearly so, at sides of twig, petals yellowish. Fruit elliptic, flattened, bluish berry on curved stalk, shooting sticky seed suddenly with force to about 6 m.

Juneau south to Dixon Entrance. Common. Flowers August-September; fruits the following September. Forms "witches brooms" on branches of hemlock; considered a serious tree disease which significantly reduces growth of host tree. Its ecological role could be to thin out hemlocks in the forest. Also known as *A. tsugense* since it affects western and mountain hemlock (rarely lodgepole pine).

MENYANTHACEAE
BUCKBEAN FAMILY

Fauria crista-galli   Deer-cabbage

Hairless perennial 10-50 cm tall from thick, fleshy, reddish brown rhizome covered with remains of old leaves. Leaves basal, thick, heart- to kidney-shaped, margins with fine round teeth. Flowers few to several on long naked stalk, petals, white, wheel-shaped, mid vein and margins with wavy edges. Capsule elongated, much longer than calyx.

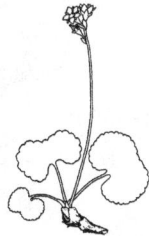

Southeastern Alaska west to Aleutians. Bogs, swamps, wet meadows, seeps and open forest. Common. Flowers late June-August. Flowers have odor of bad cheese; attracting insect pollinators. Leaves eaten by deer in summer; makes steep alpine and subalpine slopes treacherous when wet. Also known as *Nephrophyllidium crista-galli.* Deer-cabbage and buckbean related to gentians.

Menyanthes trifoliata Buckbean

Aquatic to semi-aquatic submerged perennial from thick scaly rhizome covered with old leaf bases. Leaves alternate, divided into 3 elliptic leaflets. Raceme leafless, corolla funnel-shaped, star-like petal lobes with dense beard of white hairs on inside, tube slightly longer than calyx, petals white to pink, lobes usually tinged purplish at apex. Capsules ellipsoid.

Most of Alaska. Swamps, bogs, fens, ponds and wet meadows. Common. Flowers June-August. Flowers have rank odor. Rhizomes used for emergency food by drying, grounding and washing to leach out bitterness; can be made into unpalatable flour. Used to treat scurvy, stomach ailments, relieve fever and headaches, promote hunger and get rid of intestinal worms. Birds eat the fruit.

MYRICACEAE
BAYBERRY FAMILY

Myrica gale Sweet Gale

Low aromatic shrub to 1.5 m tall branching loosely at base. Leaves spatula-shaped with few teeth at apex, finely hairy on both sides, dotted with yellow waxy glands, twigs finely hairy when young, dark brown to gray with yellow resin dots and white dots resembling lenticels. Flowers dioecious, small, inconspicuous, yellowish in spikes before leaves emerge, dotted with waxy yellow glands, petals yellow. Fruit green, 2-winged, resinous, waxy nutlet.

Southeastern Alaska west to Aleutians north to 67th parallel. Shallow water, wet coastal meadows, bogs, rocks along coast and tidal flats, mostly at low elevation. Occasional. Flowers May-June. Tea made from leaves used as a wash for boils and pimples. Leaves produce a golden yellow dye. Improves wetland soils by fixing nitrogen.

NYMPHAEACEAE
WATER LILY FAMILY

Brasenia schreberi   Watershield

Aquatic perennial covered with gelatinous sheath (except for upper surface of leaves) from slender rhizome. Leaves alternate, elliptic, floating on water surface. Sepals and petals 3 sometimes 4, purplish, small and inconspicuous, stamens numerous, flattened filaments, 1 pistil. Fruit 1-2 seeded non-splitting leathery follicle.

Known from extreme southeastern portion of Alaska (Gravina Island). Shallow ponds and sluggish streams. Young leaves edible. May have anti-algal and anti-bacterial properties useful for natural weed control. Flowers August.

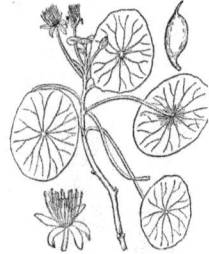

Nuphar luteum   Yellow Pond-Lily

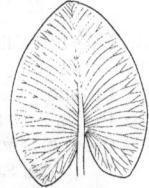

Coarse aquatic perennial from thick, submerged huge rhizome. Leaves leathery, alternate, appearing spirally arranged. Flower solitary on long stalks, arising from rhizome, sepals yellow and tinged green or yellow, petals hidden by stamens, yellowish to purple. Fruit large capsule becoming leathery and podlike, containing many large seeds.

Southeastern Alaska west to Aleutians north to 68th parallel, rarely northward. Ponds and lakes. Common. Flowers June-July. Ripe seeds roasted or ground into flour or popped like corn; rhizomes used as a starchy vegetable or ground into meal. Used to cure tuberculosus and rheumatism; rhizomes used as a blood tonic. Also known as *N. polysepalum*.

Nymphaea tetragona   Pygmy Waterlily

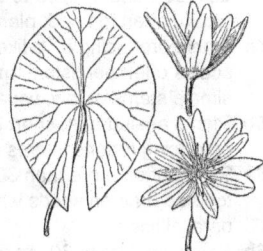

Aquatic perennial with flowers and leaves from thick submerged rhizome. Leaves ovate to arrowhead-shaped with backward pointing lobes, appearing spirally arranged. Flowers with quadrangular base, petals almost equal to sepals, white, sometimes pinkish with crimson lines. Fruit egg-shaped 1-2 cm long capsule.

Central southeastern, south-central and east-central Alaska. Ponds and lakes. Occasional. Flowers June-August.

ONAGRACEAE
EVENING-PRIMROSE FAMILY

Circaea alpina　　Enchanter's Nightshade

Perennial 10-50 cm tall from tuberous rhizome, plant delicate, spread by runners and nodes which only live for one year, usually occurs in colonies. Leaves thin, heart-shaped, hairless, ovate, acute, irregular and widely spaced, shallow teeth on margins. Petioles narrowly winged, 2 reflexed sepals, petals white to pinkish, bilobed to about middle, stigma entire. Fruit oblong, 1-celled and covered with soft bristles.

Southeastern Alaska west to Kodiak Island and Alaska Peninsula, less common in interior Alaska. Stream sides, wet areas in forests, along seeps and under *Alnus sinuata* along beaches, low to mid elevation. Associated with Devils Club on loess and other fertile soils especially on islands in the vicinity of the Stikine delta. Common. Flowers June-August.

KEY TO SPECIES OF *EPILOBIUM:*

1a. Petals (8)10-30 mm long, rounded apically, spreading; hypanthium not extended beyond ovary (calyx cleft to top of ovary). **(2)**

1b. Petals 2-10 mm long (longer in *E. luteum*), notched apically, ascending; hypanthium extending beyond ovary (calyx not cleft to top of ovary) **(3)**

　　2a. Styles pubescent at base; floral bracts much reduced; herbage hairless or nearly so; inflorescence usually with 15-more flowers.
　　　　　　　　　　　　　　　　　　　　　　　　E. angustifolium

　　2b. Styles hairless; floral bracts similar to leaves, through smaller; herbage usually minutely pubescent; inflorescence usually with fewer than 10 flowers.　　　　　　　　　　　　　　　　*E. latifolium*

3a. Petals yellow, 12-19 mm long; plants 13-80 cm tall; occurring in coastal southern Alaska.　　　　　　　　　　　　　　　　　　　　*E. luteum*

3b. Petals pinkish to purplish or white, mostly less than 12 mm long; plants of various distribution. **(4)**

4a. Leaves linear to narrowly lance-shaped, mostly 8(12) mm broad or less, entire or nearly so; plants often grayish-hairy.　　　　　　　*E. palustre*

4b. Leaves lance-shaped to oblong or ovate-lance-shaped, often more than 8 mm broad, often toothed; plants variously pubescent to hairless. **(5)**

5a. Plants producing bulb-like offsets (turions) at base of stem, fleshy overlapping scales often persistent on base of current stem, either robust herbs with mostly simple stems or low, usually much branched herbs.　　*E. ciliatum*

5b. Plants arising from short to elongate rhizomes or from taproots or fibrous roots, not producing turions; stems simple or branched. **(6)**

6a. Stems often more than 30 cm tall; rhizomes short or lacking; leaves often finely toothed; hair on seeds white (sometimes dingy); seeds minutely roughened in parallel lines.　　　　　　　　　　　　　　　　*E. ciliatum*

6b. Stems seldom to 30 cm tall; rhizomes mostly well developed; leaves somewhat entire; hair on seeds dingy; seeds smooth to somewhat roughened. *E. hornemanii*

Epilobium hornemanii   Alpine Willow-herb

Perennial 10-50 cm tall, often reddish with prostrate stem, basal leafy shoots. Leaves opposite, mostly in 2-3 pairs on short petioles, blade oblong to narrowly elliptic. Flowers small, few to several, often nodding, petals reddish-violet to pink, 4-notched. Pod-like capsules 4-10 cm long, stalked, seeds hairy.

Southeastern Alaska west to Aleutians, interior and western Alaska. Seeps, wet open areas and on rocks in alpine meadows. Common. Flowers June-August. A complex of species including *E. anagallidifolium*, *E. serulatum*, *E. alpinum*, *E. nutens* and *E. behringianum*.

Epilobium angustifolium   Fireweed

Perennial to 3 m tall with stems often purplish, tall, unbranched, densely leafy from woody rhizomes. Leaves alternate, lance-shaped, acute, hairless, paler and distinctly veined below. Flowers in long terminal racemes, flowering from base, pink-purple, rarely white, large clawed petals, styles longer than stamens, sepals hairy and reddish. Capsules covered with short, soft hairs, seeds white and fluffy.

Throughout Alaska. Open areas, meadows, beach fringe and burned areas. Common. Flowers July-August. Marrow or sweet inner pith scraped and eaten; young tender shoots good source of vitamins A and C. Leaves used for medicinal tea for stomach aches and as a sedative. Down makes good fire starter and can be combined with wool, cotton or fur. "Skin" of stem peeled, dried, soaked in water and twisted into twine for fishing nets. Flowers used to waterproof rawhide and mittens and nectar for honey. The species name means narrow leaves. Some botanists place this species in the genus *Chamerion* based on several morphological distinctions including spiral rather than opposite or whorled leaf arrangement, absence rather than presence of a hypanthium and sub-equal stamens rather than stamens in two unequal whorls. This species also known as the "Sourdough Calendar" since as the story goes: when the fireweed finishes flowering, it will be six weeks until the snow comes to Southeast Alaska.

123

Epilobium ciliatum Glandular Willow-herb

Perennial 15-150 cm tall from tap or fibrous roots or rhizomes, turions present, stems sometimes purplish, pubescence often extending down stem in lines below leaf bases. Leaves opposite, sometimes becoming alternate above, narrowly lance-shaped to oval or elliptic with minute teeth, stalkless and somewhat clasping. Flowers small, sepals oval-lance-shaped, glandular-pubescent, petals pink to purplish. Seeds raggedly roughened in parallel lines, hair dingy or white.

Southeastern Alaska west to Aleutians, scattered locations in Interior. Moist sites in woods, thickets, meadows, roadsides and along beaches. Fairly common. Flowers late June-September. *E. leptocarpum* closely related but smaller. Also known as *E. glandulosum, E. watsonii* and *E. adenocaulon.*

Epilobium latifolium River Beauty

Low growing perennial 5-40 cm tall arising from woody base, lacking rhizomes, stem sometimes purplish. Leaves smooth, opposite (or whorled) below, usually alternate above, fleshy, hairless, elliptic-oval to lance-shaped, entire or toothed margins, not veiny. Flowers in upper axils, style hairless, shorter than stamens, 4 broad petals with long narrow tube, bright pink to pink-purple, rarely white. Capsules usually purplish, seeds smooth, hair dingy.

Most of Alaska. River bars, along streams, scree slopes, open gravelly areas, alpine meadows and roadsides. Common. Flowers June-August. Young leaves and inner stem eaten raw or boiled.

Epilobium luteum Yellow Willow-herb

Perennial 20-80 cm tall with simple stems or branched from near base from stout rhizomes. Basal leaves small, soon withering; stem leaves numerous, opposite, overlapping, almost stalkless, lance-shaped to elliptic. Petal margin gently wavy, styles hairless, longer than yellow petals. Capsules glandular-hairy, seeds smooth, hairs dingy.

Southeastern Alaska west to Aleutians. Moist sites along streams, springs and lakes. Uncommon. Flowers late July to early September.

124

Epilobium palustre   Swamp Willow-herb

Perennial herb 10-40 cm tall, stem sometimes reddish, erect with short soft hairs above, simple or branched, plant with long thread-like runners at base ending with bud, producing turions. Leaves opposite below, alternate above, stalkless, linear-lance-shaped, entire margins. Flowers small, drooping when young, petals notched, pink or whitish. Pod short-hairy especially along margins, seeds covered with small, nipple-shaped bumps.

Most of Alaska. Wet places, along rivers, bogs and meadows. Common. Flowers July-August.

OROBANCHACEAE
BROOMRAPE FAMILY

Boschniakia rossica   Ground-cone

Cone-like parasitic herb to 40 cm tall and 2.5 cm in diameter from short, fleshy, bulblike underground stem. Leaves thick scales, triangular to ovate, yellowish to purplish with entire to minutely shredded margins. Many-flowered with one 3-lobed lip flower per leaf axil, less than 2.5 cm long on thick stalk, flower bracts hairy fringed, petals purplish to yellow; fruit capsule, seeds numerous, tiny.

Widely distributed in Alaska. Parasitic mostly on roots of *Alnus* but other trees and shrubs as well, along trails, in forests and along beaches. Locally common. Flowers July-August. Eaten in fruiting stage by some indigenous people. Brown bears feed on ground-cone. Some studies suggest this species encourages strong free radical scavenging activity and consequently has inhibitory effects on the disorders caused by free radical production in living tissue.

Boschniakia hookeri (Hookers Ground-cone) Yellow to dark red or purple; differs from *B. rossica* in having scale leaves with entire margins, pubescent filaments, and anthers much exceeding connective. Reported from northern British Columbia and may be in southern southeastern Alaska as a parasite on roots of *Gaultheria shallon*.

PLANTAGINACEAE
PLANTAIN FAMILY

Plantago macrocarpa   Seashore Plantain

Hairless, stout, perennial herb 10-60 cm tall from thick, stout, vertical root. Leaves elliptic to lance-shaped, several-nerved, erect. Flowers greenish to brownish, inconspicuous in 1-several scapes, inflorescence short and dense at peak flowering, elongating in fruit, sepals dark-brown, ovate to elliptic, 4 protruding anthers. Capsules large, elliptic, non-splitting.

Southeastern Alaska west to Aleutians. Wet places, beach meadows, brackish marshes and muskegs. Common. Flowers May-June. Tender young leaves eaten raw in salads or cooked like spinach.

Plantago maritima   Goose-tongue

Perennial herb 5-25 cm tall. Many long, narrow leaves arising from base of plant. Central flower stalk rising barely above leaves with dense, blunt spike of small flowers with four greenish or white petals. Capsules 3-4 mm long, seeds 2-4.

Southeastern Alaska west to Aleutians where scattered. Beaches. Common. Flowers late May-August. Leaves edible when young and tender. This plant can be confused with and often grows in the same habitat as *Triglochin maritimum* which contains cyanide-producing glycosides that can be poisonous and deadly to livestock. Goose-tongue has flower stalks barely as long as the leaves and indentations on the leaves. *Triglochin* at maturity has flowers much longer and above the leaves.

POLEMONIACEAE
PHLOX FAMILY

Polemonium caeruleum   Blue Jacobs-ladder

Perennial herb 20-100 cm tall, hairy especially above, stem solitary. Leaves primarily basal, pinnately compound, leaflets 10-25 mm long and 3-10 mm broad, lance-shaped to elliptic. Flowers bell-shaped with 5 sharp pointed calyx lobes, corolla hairy-margined, lobes about twice as long as tubes, petals blue to violet. Fruit globular capsule.

Haines area, west and north of Yakutat throughout Alaska. Wet meadows and along streams. Common. Flowers June-July. Highly variable species; also known as *P. acutiflorum*.

Polemonium boreale (Northern Jacob's-ladder) 15-20 cm tall, stem solitary or few together from rhizome, apex leaves pinnately compound, reduced upwards with shorter (4-12 mm long) and narrower (1-5 mm broad) leaflets; corolla violet to blue, rarely white, lobes slightly longer than tubes. Tundra and woodlands widely distributed in Alaska from Haines northward.

POLYGONACEAE
BUCKWHEAT FAMILY

Koenigia islandica   Koenigia

Hairless diminutive annual herb 2-15 cm long from slender taproots, more or less reddish. Leaves elliptic to ovate, blunt. Flowers single or in small heads, calyx with 3(4) lobes and as many stamens, petals greenish, whitish, or reddish. Achenes hairless, enclosed by perianth.

Lynn Canal, Prince William Sound west to Kodiak Island, Alaska Peninsula, and eastern and western Aleutians. Wet places on bare soil, snow beds. Widely scattered sites; often overlooked.

Oxyria digynia Mountain Sorrel

Perennial herb 5-60 cm tall from fleshy taproot, vegetative parts often reddish tinged, stems simple. Leaves mostly basal, 1-2 kidney-shaped wavy-margined leaves on each stem. Flowers small, clustered along ribbed stalk, rising above leaves, stalks often branched, petals greenish to crimson. Achenes flattened, prominently veined.

Most of Alaska. Wet open areas, snow beds, alpine. Common. Flowers June-July. Leaves but not roots eaten raw. Contains oxalic acid.

Polygonum viviparum Alpine Bistort

Erect herb from thick, hard, usually contorted rhizome, stems with fine grooves. Basal leaves well developed, blades narrowly oblong; stem leaves reduced upwards. Flowers in spike with lower flowers developing into bulbets, sometimes growing into small plants on stem, calyx of upper flower 5-parted, petals white or pink. Achenes 3-angled, pale brownish, lustrous, seldom developing.

Most of Alaska. Bogs and meadows and along rivers from sea level to alpine. Common. Flowers late June-August. Rhizome collected in early spring and eaten raw; tastes like almonds.

Polygonum amphibium (Water Smartweed) Floating when aquatic, 50-100 cm long or longer, leaves narrowly elliptic to oblong with smooth margins; flowers in 1-2 spike-like panicles, perianth bright pink, lobes almost equal; dark brown lens-shaped achenes. Ponds, streams, and mud banks from southeastern Alaska west to Prince William Sound and central western Alaska. Flowers July-August.

Polygonum aviculare (Prostrate Knotweed) Annual herb, stems grooved and usually triangular at least when young, 10-100 cm long, leaves elliptic to lance-shaped, reduced upwards; 1-5 flowers in axils, perianth united about half to one-third length, 5-lobed, lobes greenish with white or pink edges, lobes unequal; achenes 3-angled, brown. Weedy species of disturbed sites throughout southern Alaska. *P. arenastrum* which has more uniform sized leaves and less deeply divided perianth and *P. boreale* are closely related. This species has many medicial uses.

128

Polygonum caurianum (Alaska Knotweed) Annual, stems grooved and often triangular, 5-50 cm long, leaves oblong to elliptic; flowers in axils, 5 perianth segments greenish with white to red margins, outer ones somewhat broader than inner; achenes 3-angled, straw colored to brownish. Waste places, beaches, dunes, roadsides and gravel bars in much of Alaska south of Brooks Range.

Polygonum fowleri (Fowler Knotweed) Stems grooved, circular in transverse section, 10-50 cm long or more, leaves not as crowded as in other species, elliptic to oblanceolate; flowers 1-4 clustered in axils, perianth united about one half their length, lobes greenish with pink margins; achenes 3-angled, yellowish brown, protruding. Coastal areas from southeastern Alaska west to Kodiak Island and Kenai Peninsula. Flowers July-August.

Rumex occidentalis Western Dock

Perennial herb from taproots to 2 m tall, usually unbranched below flower, often reddish tinged. Leaves basal, long-petiolated, lance-ovate with heart-shaped base, upper leaves smaller. Flowers in dense panicle with leafy bracts near base, petals greenish. Achenes brown, lustrous, fruiting pedicel obscurely jointed near or below middle.

Southeastern Alaska west to Aleutians, less common in Interior. Wet open areas and beaches. Flowers July-August. Young stalks, stems and leaves edible; seeds ground into meal and yellow dye made from roots. Also known as ***R. aquaticus.***

Rumex salicifolius Beach Dock

Herb 20-60 cm tall, stems rarely erect, branching from lower nodes. Leaves mostly on stems, narrowly lance-like or linear, sometimes wavy. Flowers numerous in panicles with leafy bracts mostly near base, petals greenish-brown to deep pink. Fruit coarsely wrinkled-pitted with uneven edges, achenes smooth.

Southeastern Alaska west to Aleutians north to 66th parallel. Disturbed soils, beaches and river banks. Occasional. Flowers July-August. Also known as ***R. transitorius***.

PORTULACACEAE
PURSLANE FAMILY

Montia parvifolia   Little Leaf Montia

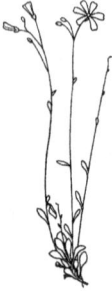

Perennial herb 10-30 cm tall from elongate rhizomes with elongate stolons. Basal leaves somewhat sheathing basally, blades oval to round, stem leaves alternate, mostly smaller and narrower than basal leaves. Inflorescence raceme-like, erect, terminal, lower flower subtended by bract, 2 rounded white to pink sepals, 5 stamens. Capsules egg-shaped, seeds lustrous black.

Southeastern Alaska. Moist rocky outcrops along streams and beaches. Occasional. Flowers June-August. Also known as *Claytonia parvifolia*.

Claytonia sibirica   Siberian Spring-beauty

Annual or rarely perennial herb 10-40 cm tall from fibrous roots or slender rhizome. Basal leaves long-stemmed, lance-shaped to ovate, round or elliptic; stem leaves 2, opposite, ovate to broadly lance-shaped. Flowering stems to 50 cm long, prostrate to erect, flowers terminal and in leaf axils, many nodding in bud, spreading in flower, 5 white to pink notched petals basally united, 2 sepals. Capsules with black and lustrous seeds.

Southeastern Alaska west to Aleutians. Moist shady places and along coast. Common. Flowers May-early September. Leaves eaten raw, steamed, or used for tea; potato-like subterranean bulb-like stem edible. Also known as *Montia sibirica*.

Claytonia tuberosa (Tuberous Spring-beauty) 6-15 cm tall from short, thick subterranean bulb-like corm, stems mostly solitary, arising from corm, usually whitish, basal leaves 1-2, lance-shaped to linear lance-shaped, stem leaves narrowly elliptic or narrowly oblong; sepals 2, ovate to round, petals white or rarely pink, basally united; seeds round and black or brown. Moist sites in tundra, on islands of Bering Sea and Seward Peninsula eastward, also at mouth of Stikine River. Subterranean bulb-like corm eaten boiled or roasted; leaves used for salad.

Montia chamissoi (Toad-lily) 5-25 cm¯ tall, from slender rhizomes and stolons which produce bulb-like offsets, flower stems often branched, basal leaves reduced or lacking, stem leaves opposite, oblanceolate to elliptic, tapering to petiole or stalkless; flowers terminal, in axils of inflorescence, single bract at base of lowest pedicel, 2 sepals, 5 white or pinkish petals, stamens mostly 5. Marshes, seeps, springs, or streams on coast and islands of southern Alaska and less commonly in interior southern Alaska. Flowers June-August. Also placed in genus *Claytonia*.

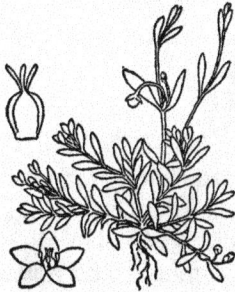

Montia fontana (Water Blinks) Small slender annual, prostrate to ascending, 3-25 cm long, often much-branched, basal leaves reduced or much-branched, stem leaves opposite and oblanceolate to elliptic; inflorescence nodding, raceme-like, terminal, petals white, united about half their length, stamens usually 3; black seeds. Wet soil or shallow water on coast and islands in southern and western Alaska. Flowers May-June.

PRIMULACEAE
PRIMROSE FAMILY

KEY TO GENERA OF *PRIMULACEAE:*

1a. Leaves along stem; flowers axillary or in axillary inflorescences. **(2)**

1b. Leaves all basal; flowers terminal or in terminal inflorescences. **(4)**

    2a. Corolla lacking; plants of saline soils.   *Glaux*

    2b. Corolla present; plants usually not of saline soils. **(3)**

    3a. Flowers sulfur yellow, in dense axillary clusters; plants tall, semi-aquatic.
        Lysimachia

    3b. Flowers white or pinkish, solitary on long peduncles; plants of various habitats, seldom semi-aquatic.   *Trientalis*

4a. Corolla lobes distinctly reflexed; flowers nodding.   *Dodecatheon*

4b. Corolla lobes spreading to erect, not reflexed; flowers erect or spreading. **(5)**

5a. Plants densely tufted perennials with persistent, densely overlapping, narrow leaves; peduncles solitary, each with single flower; corolla pink.   *Douglasia*

5b. Plants various, but seldom densely tufted; peduncles 1-more, each with 1-few flowers; corolla white to pink. **(6)**

6a. Flowers constricted at throat, white and borne in umbels (rarely solitary), or sometimes fading pinkish; corolla tube usually shorter than calyx.   *Androsace*

6b. Flowers open at throat, white, or more commonly pink, inflorescence with umbels, commonly with 2-more flowers; corolla tube usually longer than calyx.   *Primula*

Androsace septentrionalis   Pygmyflower

Annual herb 6-40 cm tall from taproot. Leaves numerous in dense basal rosette, linear-lance-shaped to oblanceolate with small branched hairs. Flower scapes of varying lengths emerging from long single stem with linear-lance-shaped bracts subtending inflorescence, calyx lobes narrowly triangular, corolla tubular, petals white. Capsules 2-4 mm long.

Throughout Alaska. Gravelly and sandy soils. More common in Interior. Flowers late June-July.

Androsace lehmanniana (Sweet-flowered Androsace) Forms mats usually less than 10 cm tall with woody base and prostrate stems each with terminal rosette; umbels subtended by few to several pouch-like bracts, corolla white to cream, fading pinkish with yellowish center. Arctic and alpine tundra, less common in woods in most of Alaska; found on William Henry Mt. in southeastern Alaska. Also known as *A. chamaejasme.*

Dodecatheon jeffreyi Jeffrey's Shooting-star

Perennial 10-60 cm tall from slender pale rhizomes. Leaves lance-shaped, somewhat acute to obtuse, margin irregularly rounded, blunt toothed or entire. Petal lobes 10-25 mm long, face of petals purple-black surrounded by yellow to white, connective tube black (rarely yellow), staminal filaments short (less than 1.5 mm long) or obsolete, rough, stigma conspicuously enlarged, petals magenta to lavender or white. Fruit an egg-shaped capsule.

Coastal southeastern and south-central Alaska. Wet meadows and bogs from sea level to alpine. Common. Flowers June-July.

Dodecatheon pulchellum Pretty Shooting-star

Perennial with pale root, 5-50 cm tall, hairless from thick erect woody base. Leaves round with toothed to entire margin, thick, oblong-lance-to spoon-shaped, gradually narrowed to long winged petiole. Petal lobes shorter than those of *D. jeffreyi*, 2.5-12 mm long, face of petals with yellow zigzag around connective, filaments united into orange tube, anthers purplish on outside, staminal filaments 1 mm long or longer, petals magenta to lavender. Fruit an egg-shaped-cylindric capsule.

Southeast Alaska west to NE Aleutians and Kodiak Island and Interior. Open areas, bogs and coastal meadows, mostly at low elevation. Common in SE Alaska, less common north. Flowers May-early July. Also known as *D. pauciflorum.*

132

Glaux maritima   Sea Milk-wort

Succulent, hairless perennial herb 3-25 cm tall from short rhizomes with fibrous or tuberous roots.   Leaves oval to narrowly oblong, opposite below, almost opposite or alternate above, jointed to stem. Flowers solitary in axils near middle of stem, calyx cup-shaped, 5 petal-like white or pinkish lobes equaling or longer than tube, petals lacking. Capsule 2-3 mm long, almost spherical, few-seeded.

Coastal southeastern and south-central Alaska. Salt marshes, along beaches, tide flats, saline marshes and meadows, low elevation. Common. Flowers late June-early September. Tea brewed from Sea Milk-wort given to nursing mothers to increase milk supply. Fleshy rhizome eaten boiled but too much induces sleepiness and a sick feeling.

Lysimachia thyrsiflora   Tufted Loosestrife

Hairless, perennial to 80 cm tall from long, thick creeping rhizome. Lower leaves scale-like, upper leaves opposite, lance-shaped to narrowly elliptic, stalkless and dotted with black glands. Flowers dense in leaf axils, stamens protruding, petals yellow with purple spots or streaks. Capsule ovate, seeds, few, pitted.

Southeastern Alaska west to Aleutians, also in scattered sites in interior southwestern to east-central Alaska. Wet marshes and lake shores low to mid elevation. Uncommon. Flowers June-July. Used to deter gnats and flies. Used medicinally in Asia to combat high blood pressure.

Primula cuneifolia   Wedge-leaf Primrose

Perennial 2-20 cm tall growing in patches from fibrous roots. Leaves wedge-shaped with 5-11 apical teeth. Few-flowered, petals wedge-shaped, deeply 2-lobed, pink to rose to white. Capsule almost spherical, shorter than calyx, seeds angled.

Southeastern Alaska west to Aleutians, north to Kotzebue Sound, less common in interior Alaska. Wet meadows, alpine. Fairly common on alpine ridges. Flowers June-July. *Primula eximia* (Chukch Primrose) found at Sitka and Prince William Sound.

Primula egaliksensis (Greenland Primrose) 1-27 cm tall, leaves oval or flattened spoon-shaped, margins entire or wavy, with slender petiole, bracts lance-shaped, pouch-like at base; umbels 1-9-flowered, calyx teeth shorter than tube, corolla white or violet, tube longer than calyx; seeds smooth. Moist soils, stream banks, marshes and rocky outcrops in widely disjunct sites throughout Alaska. Flowers June-July.

Primula stricta (Hornemann Primrose) 1.5-30 cm tall, leaves inversely lance-shaped, margins entire or with irregular sharp forward pointing teeth, not or only slightly mealy below, broadly petiolated, bracts pouch-like or swollen at base; umbels 2-8-flowered, corolla pink purple, tube longer than calyx; seeds angled. Moist sites in southeastern, western and northern Alaska. Poorly understood species.

Trientalis arctica   Arctic Starflower

Simple, erect, perennial 5-25 cm tall arising from slender rhizomes. Leaves simple, entire, alternate below a crowded whorl, oval-elliptic to broadly lance-shaped in whorl of 5-6 at top of stem. Flowers 1-3 on long slender red stalks, petals united at base, mostly seven sepals and petals, sepals sharp-pointed, as long as petals, petals white or pinkish white. Fruit capsule that splits into 5 parts.

Southeastern Alaska west to Aleutians and north to 67th parallel. Forests, beach forests, open areas, bogs and subalpine meadows. Common. Flowers May-August. Also known as ***T. europaea***.

PYROLACEAE
WINTERGREEN FAMILY

Chimophila umbellata   Pipsissewa

Evergreen sub-shrub 10-30 cm tall, stout, slightly woody perennial with creeping rhizome, stems hairless, greenish. Leaves whorled, lance-shaped, bright green and lustrous above, short petioled, hairless, sharply toothed. Flowers 4 to 15, nodding in cluster at end of twigs, saucer-shaped, petals separate, usually 5, reddish to pink. Fruit spherical, dry, 5-parted, many seeded capsule 6-8 mm in diameter.

Found at Hyder, in forests around Haines and in Taku River area in southeastern Alaska. Forests at low to mid-elevation. Uncommon. Flowers June-July. Leaves put in bath for sore muscles; tea used for colds and flu.

Hypopitys monotropa   American Pinesap

Fleshy saprophyte to 30 cm tall, unbranched, yellowish to pinkish, drying black, usually pubescent, arising from dense mass of fleshy roots. Leaf margins entire to fringed with coarse hairs, scale-like, lance-shaped to linear-lance-shaped, thick, crowded near stem base. Flowers in leaf axils and bracted terminal raceme, nodding before and after flowering, erect in fruit, terminal flowers largest and usually with five petals, lower ones mostly with four petals, petals much longer than sepals and overlapping one another, petals waxy-white to yellowish. Fruit an almost spherical capsule.

Southeastern Alaska. Dense forests. Occasional. Flowers July-August.   Name comes from Greek for beneath (hypo) and pine-tree (pitys) referring to its preferred habitat. Also known as *Monotropa hypopitys*.

Monotropa uniflora   Indian-pipe

Saprophyte 5-25 cm tall, waxy-white or pinkish perennial blackening with age with cluster of flowering stems arising from dense rounded mass of matted mycorrhizal roots. Leaves scale-like, linear or lance-shaped to oval. Flowers solitary, nodding, narrowly bell-shaped, waxy, calyx of 4 irregular bract-like sepals or lacking, petals usually 5, longer than sepals, petals waxy white, drying black. Fruit erect almost spherical capsule about 6 mm long.

East and south of Ketchikan on mainland and Admiralty Island. Dense forests. Uncommon. Flowers July-August. Its name comes from Greek for one (monos) and direction (tropos) due to its having flowers turned to one side.

KEY TO SPECIES OF *PYROLA:*
1a. Styles straight or nearly so; pores of anthers stalkless; stigma usually much broader than style. **(2)**
1b. Styles bent or curved; pores of anthers usually borne on short tubes (stalkless or nearly so in *P. grandiflora*); stigma only slightly broader than styles. **(3)**
　2a. Styles 2 mm long or less, not (or seldom) protruding from flower; flowers not grouped on one side of stem; petals pinkish to cream. *P. minor*
　2b. Styles over 2 mm long, protruding from flower; flowers grouped on one side of stem; petals greenish white.　　　*P. secunda*
3a. Flowers pink to purplish; sepals longer than broad.　　*P. asarifolia*
3b. Flowers pale, greenish yellow, or creamy white to pinkish; sepals various. **(4)**
4a. Sepals broader than long; flowers greenish yellow.　　*P. virens*
4b. Sepals longer than broad; flowers creamy white to pinkish.　　*P. grandiflora*

Pyrola asarifolia   Large Liverleaf Wintergreen

Perennial evergreen 10-40 cm tall from long, creeping, branched rhizome; flowering stem with 1-3 rough-textured bracts below inflorescence. Leaves large, rounded, leathery in basal cluster, green above, usually purplish below. Flowers nodding along elongated raceme with style off to one side and flaring collar below stigma, anthers pink, reddish, or yellow in age, 5 distinct petals, pale pink to dark red. Capsule 6-8 mm long.

Southeastern Alaska west to Aleutians north to 68th parallel. Forest, gravelly open areas, beach forest and recently deglaciated areas. Locally common. Flowers June-July. Used as a poultice for sores and swelling.

Pyrola chlorantha   Greenish Wintergreen

Perennial herb 10-30 cm tall with long slender rhizomes and leafy sterile shoots, flowering stem usually single. Leaves round to broadly-oval, rounded to obtuse at apex, leathery, entire or finely blunt-toothed, usually exceeded by petioles. Flowers in racemes with few-10 flowers, sepals triangular, style curved, 5 petals, pale yellowish or greenish-white and veined with green. Fruit depressed-spherical capsule.

Taku Inlet, Juneau-Lynn Canal area to south-central and continental southeastern Alaska. Mossy coniferous forests. Common. Flowers June to August. Also known as *P. virens.*

Pyrola minor   Lesser Wintergreen

Perennial herb 10-25 cm tall from slender rhizomes, flowering stem in most cases single, naked, or with one or 2 rough-textured bracts. Leaves basal, broadly elliptic or round, dark green and rather thin, margins with fine, rounded teeth. Flowers in 2-few flowered crowded raceme, sepals pinkish, 5 oval white or pink petals, anthers 10, ovate with terminal pores and flattened filaments, style straight, stigma 5-lobed. Fruit depressed-spherical capsule.

Southeastern Alaska west to Aleutians north to 66th parallel. Dense mossy coniferous forests and meadows. Uncommon. Flowers July-August.

Pyrola secunda   One-sided Wintergreen

Perennial evergreen to 20 cm tall from slender, much-branched rhizome, stem usually single, leafy toward base and often woody with 1-2 rough-textured bracts. Leaves numerous, oval to elliptic, margins with fine rounded to minute teeth, rounded to obtuse at base, acute at apex, dark green above, paler beneath. Flowers many in one-sided rather dense racemes, erect in bud, drooping at peak flowering, sepals triangular, 5 broadly oval light greenish to white petals, each with 2 small tubers on inner surface at base, anthers without tubes, pores large. Fruit depressed-spherical capsule to 5 mm in diameter.

Most of Alaska. Open mossy coniferous forests, low to subalpine elevation. Occasional to common. Flowers June-August. Also known as *Orthilia secunda*. Pyrolas are called wintergreen because they are evergreen; not the source of "oil of wintergreen" which comes from a relative of salal (*Gaultheria shallon*). Leaves used for skin problems including insect bites.

Pyrola uniflora   Single Delight

Low growing, delicate perennial from slender rhizome with leafless scape to 17 cm tall, usually with 1-2 bracts about mid-stem. Leaves oval with small sharp teeth in basal rosette. Flowers single, nodding, 5 long, thin and whitish sepals, 5 reflexed ovate waxy white to pale pink petals, 10 stamens. Fruit round erect capsule to 8 mm broad, tiny numerous seeds.

Southeastern Alaska west to Kenai, northern Alaska Peninsula, Kodiak Island and north throughout Alaska. Dense forests on rotten wood or organic soils at low elevation. Common. Flowers June-August. Whole plant brewed for tea to use as cold remedy for stomach disorders, lung troubles, smallpox, cancer and tuberculosis. Also known as *Moneses uniflora*, Shy Maiden and Toad's Lampshade.

RANUNCULACEAE
BUTTERCUP FAMILY

KEY TO GENERA OF *RANUNCULACEAE:*
1a. Flowers distinctly irregular, mostly dark blue to purple.　　　*Aconitum*
1b. Flowers regular, seldom dark blue to purple. **(2)**

2a. Petals conspicuously spurred.　　　*Aquilegia*
2b. Petals not at all spurred. **(3)**

3a. Perianth segments all alike in color and texture (called sepals). **(4)**
3b. Perianth segments of 2 distinctive types (apparently sepals and petals), although sepals may fall off early. **(6)**

　　　4a. Leaves simple, blades kidney- to heart-shaped.　　　*Caltha*
　　　4b. Leaves compound, or if simple, then deeply lobed or dissected. **(5)**

　　　5a. Stem leaves opposite or whorled; flowers often showy.　　*Anemone*
　　　5b. Stem leaves alternate or flowers inconspicuous.　　　*Thalictrum*

6a. Flowers in terminal racemes; pistils 1; fruit fleshy red or white berry.　　*Actaea*
6b. Flowers not in terminal racemes; pistils few to numerous; fruit an achene or follicle. **(7)**

7a. Leaves 1-3 time ternately parted or dissected; fruit of distinctly stipitate follicles.
　　　　　Coptis
7b. Leaves various, but usually not as above; fruit of stalkless or almost stalkless follicles or achenes; fruit 1-seeded achene.　　　*Ranunculus*

Aconitum delphinifolium　　Mtn Monkshood

Perennial from tuberous taproots 10-120 cm tall, stems hairless below with backward facing hairs in upper parts.　Leaf blade palmately lobed, few, round in outline, hairless.　Flowers borne in terminal racemes, hood rounded, beaked, petals hooked at apex, dark blue, occasionally white. Follicles hairless to sparsely hairy.

Most of Alaska.　Meadows and along streams to alpine.　Common. Flowers July-September. Leaf form highly variable.　Whole plant extremely poisonous: contains aconitine which paralyzes nerves, lowers body temperature and blood pressure.　Northern peoples applied poison from this plant to whale spears.

Actaea rubra   Baneberry

Perennial 30-100 cm tall, stems from thick rhizome, branching above, sparsely hairy. Stem leaves large and maple-like, ternately compound with coarse teeth, basal leaves lacking. Flowers small, clustered at top of stem, stamens longer than petals, sepals whitish or purplish, petals white. Fruit shiny red (sometimes white) pea-sized berries with white pulp on short, thick stalk.

Throughout Alaska south of 66th parallel. Stream banks, forests, beaches and open slopes. Common. Flowers late May-July. **Warning**: Berries, leaves and root contain Protoanemonin found in all members of this family, which causes vomiting, bloody diarrhea, paralysis of respiration and death. As few as 6 berries can cause death in an adult.

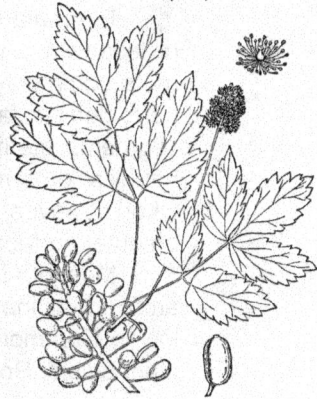

Anemone multifida   Cut-leaf Anemone

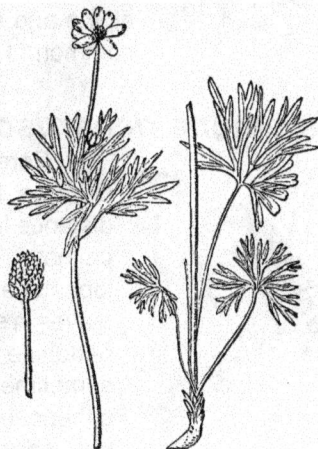

Hairy, perennial 15-60 cm tall, loosely tufted from thick, many-headed woody base, vegetative parts pubescent with long, spreading hairs. Basal leaves 2-3 times ternately divided into narrowly lance-shaped, acute lobes; stem leaves in 1-2 whorls. Flowers solitary, or 2-3, showy, sepals silky on outer surface, purple, reddish, yellow, white or bicolored. Fruit cluster of woolly egg-shaped to almost spherical achenes.

Juneau north throughout Alaska. Open woods, meadows, grassy beaches and roadsides. Common. Flowers June-July.

Anemone narcissiflora   Narcissus Anemone

Perennial herb 10-60 cm tall from woody bases, more or less with long soft hairs, vegetative parts hairless or more commonly spreading-hairy. Basal leaves 3-parted, parts cleft into fairly broad, more or less petiolated ultimate segments, stem leaves in 1 whorl. Flowers large, showy, hairless, on several long peduncles rising above upper whorl of leaves, sepals white to cream, often bluish-tinged on outer surface. Achenes hairless 6-9 mm long.

Most of Alaska. Grassy alpine areas, snowbeds in arctic and alpine tundra, mostly higher elevation, sometimes lower. Common. Flowers June-July. Upper root of spring plant edible. Roots used for treating wounds.

Anemone parviflora   Northern Anemone

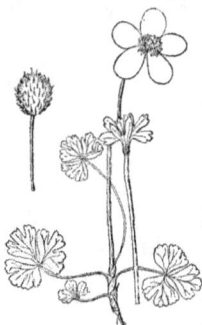

Perennial 10-30 cm tall, stems from long slender rhizomes, plant with spreading hairs or hairless. Basal leaves wedge-shaped, toothed dark-green, shiny, hairless, divided into 3-bluntly toothed lobes; stem leaves silky-hairy to hairless. Flowers solitary, large and showy, 6 silky-hairy sepals, sepals white to cream, tinged blue externally and at base. Achenes woolly hairy in spherical head.

Juneau north through most of Alaska. Meadows, alpine and arctic tundra and woods. Occasional to common. Flowers May-July.

Anemone richardsonii   Yellow Anemone

Perennial 5-15 cm tall from thin horizontal rhizome from which each leaf emerges some distance from next, stem hairy below. Basal leaves palmately divided, rounded-kidney shaped in outline, lobes shallowly divided and acutely toothed; stem leaves in single whorl. Flowers solitary, showy, one peduncle per inflorescence, sepals bright yellow sometimes bluish tinged externally. Fruiting head spherical, achenes hairless with very long beak bent backward at tip.

Juneau north throughout Alaska. Mountain meadows. Common. Flowers May-July.

Aquilegia formosa Western Columbine

Perennial herb 30-100 cm tall from stout taproot, hairless below, glandular above, branched. Leaves mainly basal, twice divided into 3's, green above, hairless below. Flowers usually 2-4 nodding, sepals and spur straight, blade yellowish, spur reddish, rarely yellow, sepals red, reddish, or rarely yellow. Follicles usually 5, more than 10 mm long, hairy with divergent styles.

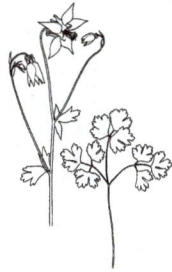

Southeastern Alaska west to Prince William Sound and Kenai Peninsula. Open areas, beaches and meadows. Common. Flowers June-July. Used medicinally for sores, diarrhea, dizziness and aching joints.

Caltha leptosepala Mountain Marsh-marigold

Perennial herb 5-30 cm tall, erect, hairless, stems single from fibrous root, stems usually with single leaf (rarely two). Leaves oblong to oval heart-shaped, rounded teeth with shallow sinus in margins, longer than broad. Flowers 1-or 2-flowered, sepals whitish often tinged with purple externally. Follicles almost stalkless with straight beak.

Southeastern Alaska west to Aleutians, less common in southern Interior Alaska. Slow running water, bogs and wet meadows, alpine. Common. Flowers June-August. Leaves and flower buds eaten cooked or raw, roots cooked as sauerkraut. This plant should be well cooked though which breaks down the poison Protoanemonin. Includes *C. biflora.*

Caltha natans Floating Marsh-marigold

Aquatic herb, floating or creeping in mud, hairless, mostly 5-50 cm long, stems leafy, with stolons rooting at nodes. Leaves kidney-shaped, margins with blunt teeth. Flowers several on stalks rising above leaves, sepals white often tinged with pink. Follicles stalkless or almost stalkless, 4-6 mm long, straight.

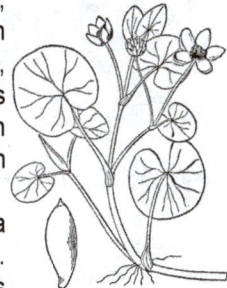

Southeastern, eastern and central Alaska disjunct to Seward and Alaska peninsulas. Occasional. Ponds and mud banks. Flowers July-August.

141

Caltha palustris   Yellow Marsh-marigold

Hairless herb, highly variable in size, stem hollow, reclining, rooting at nodes, leafy. Leaves kidney-shaped to oval heart-shaped with more or less rounded teeth in margins. One flower borne in leafy clusters on hollow stems that rise above leaves, 5-8 sepals, yellow, often tinged greenish or purplish on back. Follicles with hooked beak.

Throughout Alaska. Wet areas, in slow running water. Common. Flowers May-July. Contains poisons (anemonine and helleborin) destroyed by cooking. Leaves tasty before flowers appear; always boil and drain twice. Flower buds also edible.

Coptis asplenifolia   Fern-leaf Goldthread

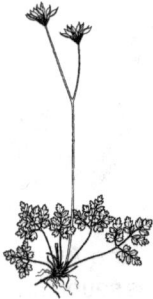

Perennial herb to 30 cm tall from gold-colored rhizomes, persistent through winter. Leaves basal, fern-like with 5-several sharply toothed segments. Flower stalk leafless, usually 1-3 delicate flowers borne on scape, sepals long, thread-like, whitish and petal-like, petals 3-8, shorter than sepals, glandular base terminated by slender blade, petals greenish-white. Follicles with short spreading tips, as many as 12 in a cluster.

Southeastern Alaska west to Prince William Sound. Forests, wet open areas and bogs low to mid elevation. Common. Flowers April-June.

Coptis trifolia   Three-leaved Goldthread

Delicate herb 5-15 cm tall from gold-colored roots. Leaves shiny, divided into three sharply toothed leaflets. Flowers single on leafless peduncle rising from base, petals reduced, sepals white and often tinged pink, petals orange at apex.

Southeastern Alaska west to Aleutians and north along western coast. Alpine tundra, moist meadows, bogs and adjacent woods. Common. Flowers late May-June. Both species have roots that can be chewed to treat mouth sores. A tea made from either species used as wash for eye irritations and sore mouth.

KEY TO SPECIES OF *RANUNCULUS:*

1a. Petals white; plants aquatic; submersed leaves divided into linear-thread-like segments. ***R. aquatilis***

1b. Petals yellow; plants terrestrial or aquatic; leaves variable, but not with linear-thread-like segments. **(2)**

2a. Leaves entire or merely rounded to sharp toothed, blades simple, not dissected; plants with conspicuous, well-developed, strawberry-like stolons. **(3)**

2b. Leaves deeply lobed or compound (at least some); plants sometimes stoloniferous, but more commonly with erect stems. **(4)**

3a. Leaves all entire, blades narrowly elliptic to lance-shaped or oblong to linear, acute at both ends; achenes less than 50, not longitudinally ribbed. ***R. flammula***

3b. Leaves with rounded to sharp teeth (at least some), blades oval, heart shaped basally, rounded to obtuse apically; achenes more than 50, longitudinally ribbed. ***R. cymbalaria***

4a. Plants aquatic or semi-aquatic; leaves often occur in two forms, at least some finely dissected; achenes hairless, beakless or beak less than 0.5 mm long. **(5)**

4b. Plants neither aquatic nor semi-aquatic with finely dissected leaves, though sometimes growing in wet places; stems mostly erect, seldom rooting at nodes; achenes hairy or beak more than 0.5 mm long. **(6)**

5a. Leaf blades with 3-5 broad, entire lobes, mostly 3-8(9) mm long; beak of achene less than 0.2 mm long. ***R. hyperboreus***

5b. Leaf blades dissected into numerous narrow lobes, mostly 10-30 mm long; beak of achenes over 0.3 mm long. ***R. gmelinii***

6a. Petals 7-16; plants hairless; stem leaves 1, or more commonly lacking; leaf blades 3-5-lobed, lobes round or blunt toothed; sepals yellow, deciduous. ***R. cooleyae***

6b. Petals usually 5; plants pubescent or hairless; stem leaves usually more than 1.**(7)**

7a. Sepals with conspicuous reddish brown or blackish hairs externally. ***R. nivalis***

7b. Sepals with yellowish or whitish hair, or hairless externally. **(8)**

8a. Leaf blades less than 3 cm long; plants rarely more than 20 cm tall; stems below inflorescence hairless or nearly so (pedicels sometimes hairy). **(9)**

8b. Leaf blades more than 3 cm long, or plants more than 20 cm tall, or stems below inflorescence long-hairy, or all of these. **(12)**

9a. Petals 1-3.5(4) mm long; plants mostly 7 cm tall or less. ***R. pygmaeus***

9b. Petals 4-18 mm long; plants often more than 7 cm tall. **(10)**

10a. Pedicels hairless (sparsely hairy in some ***R. eschscholtzii***). **(14)**

10b. Pedicels hairy. ***R. gelidus***

11a. Blades of lowermost leaves deep heart-shaped; petals almost equal to or slightly longer than sepals; beaks of achenes curved, to 0.5 mm long. ***R. verecundus***

11b. Blades of lowermost leaves ending abruptly to rounded or heart-shaped; petals distinctly longer than sepals; beaks of achenes straight, 0.8-1 mm long.
R. eschscholtzii

12a. Stems decumbent, rooting at nodes and more or less stoloniferous. ***R. repens***

12b. Stems erect or ascending, usually neither rooting at nodes nor stoloniferous. **(13)**

13a. Flowers mostly less than 15(20) mm broad. **(14)**

13b. Flowers (12)15-30 mm broad. **(15)**

14a. Cluster of achenes 3-7 mm long; beak of achenes hooked; receptacle hairless. ***R. uncinatus***

14b. Cluster of achenes 7-12 mm long; beak of achenes straight; receptacle hairy. ***R. macounii***

15a. Blades of basal leaves pinnately compound with 3-7 leaflets, leaflets all stalked; beak of achene 2.5-4 mm long. ***R. orthorhynchus***

15b. Blades of basal leaves palmately lobed or palmately compound with 3 stalkless or short-stalked leaflets. ***R. occidentalis***

Ranunculus aquatilis   Water Crowfoot

Aquatic, mostly perennial herb with weak, floating or submerged stems, rooting at lower nodes, hairless or sparsely short-hairy. Leaves along stem, alternate, dissected into thread-like segments that collapse when removed from water, blades palmately 3-5 lobed, sepals spreading to reflexed, greenish, hairless, 5 white petals. Achenes in spherical cluster.

Throughout Alaska. Ponds and streams, low to mid elevation. Occasional. Flowers June-August. Also known as *R. confervoides* and includes *R. trichophyllus*.

Ranunculus cooleyae   Cooley Buttercup

Aquatic perennial hairless herb 10-30 cm tall, not rooting at nodes. Basal leaves hairless, shiny, simple blades, 3-5 lobed, margins with round or blunt teeth or sharp forward pointing teeth, round to kidney shaped, bracts lance-shaped, 3-lobed, or lacking, no leaves on flower stem. Flowers with 5 yellow and hairless sepals, 7-16 narrow, yellow petals. Achenes 30-70 or more in hemispheric cluster, hairless, beak curved.

Southeastern Alaska west to Prince William Sound. Alpine snow banks and open areas in subalpine. Common. Flowers May-August.

Ranunculus flammula   Creeping Spearwort

Aquatic amphibious perennial herb with vegetative stems produced into strawberry-like stolons, often rooting at nodes, hairless or nearly so; flowering stems to 10 cm tall. Leaves simple, alternate, appearing basal, linear to lance-shaped or elliptic, acute at both ends and entire. Flowers with greenish deciduous sepals, 5 yellow petals. Achenes plump, in almost spherical cluster, hairless.

Most of Alaska south of Arctic Circle. Shallow ponds, wet soil or muddy banks. Flowers July. Also known as *R. reptans*.

Ranunculus occidentalis   Western Buttercup

Perennial herb 10-90 cm tall, not rooting at nodes, with spreading hairs or hairless. Basal leaves simple or 3-foliate, blades 3-parted, lobes again lobed, stem leaves alternate, bracts with three oblong lobes, stalkless. Sepals greenish, reflexed as flowers open, pubescent, deciduous, 5 (rarely more) yellow petals. Achenes in hemispheric clusters, hairless, beak slightly curved and hooked.

Southeastern, south-central and southwestern Alaska. Moist meadows and tundra, disturbed sites low to mid elevation. Common. Flowers late April-September.

Ranunculus uncinatus   Little Buttercup

Annual or perennial, erect to 1 m tall, not rooting at nodes. Mostly basal leaves 3-parted, segments lobed and toothed, stem leaves alternate. Five pubescent, deciduous, greenish, reflexed sepals, 5 small yellow petals clustered above lance-shaped leafy bracts. Achenes in hemispheric clusters, beak curved.

Coastal southeastern Alaska west to Aleutians. Moist soil in woods, thickets, meadows, beaches and along streams Common. Flowers July-August. Also known as *R. bongardii*.

Ranunculus cymbalaria (Shore Buttercup) Tufted with thread-like stolons, rooting at nodes, leaves simple, rounded to heart-shaped, margins with round or blunt teeth; flowering stems 10-30 cm tall, sepals deciduous, 5 or more yellow petals; achenes in cylindrical cluster, hairless with straight beak. Muddy sites and tide flats, along streams and in bogs disjunctly in southern and central Alaska from Juneau north to Prince William Sound. Flowers May-July.

Ranunculus eschscholtzii (Snowpatch Buttercup) Perennial with fibrous roots, 5-25 cm tall, not rooting at nodes, hairless, basal leaves simple, kidney-shaped to oval, 3-cleft and again lobed, 1-3 stem leaves, alternate or lacking, bracts usually with 3 entire lobes and stalkless; 5 sepals yellowish, hairless or with yellowish or brownish hairs, dropping off soon after flower opens, 5 yellow petals; achenes in more or less elongated cluster, hairless. Alpine tundra, meadows and talus slopes from southeastern Alaska west to Aleutians. Flowers June-July. Hybrids with *R. nivalis* with which it is closely related may occur.

Ranunculus gmelinii (Small Yellow Water-buttercup) Sparsely hairy stems prostrate or floating, 10-40 cm long, rooting at nodes, hairless or pubescent; leaves mostly on stems, alternate, 3-parted and again parted, at least some with very narrow segments, oval, deeply heart-shaped basally, sepals greenish, deciduous, 5 yellow petals longer than sepals; achenes in egg-shaped cluster, hairless, with broad backward bending beak. Mud flats, shallow ponds and marshes in most of Alaska including mainland southeastern Alaska, except coast and islands of southern Alaska. Flowers June-August.

Ranunculus hyperboreus (Farnorthern Buttercup) Stems prostrate to erect, sometimes floating, rooting at nodes, hairless, leaves mostly on stems, alternate, simple with 3-5 entire lobes; 3-5 greenish sepals, deciduous, 3-5 hairless yellow petals; achenes in spherical cluster, beak short, straight, hooked at apex. Mud flats or shallow ponds, widely distributed in Alaska. Flowers July-August.

Ranunculus macounii (Macoun Buttercup) 30-80 cm tall, erect to reclining, sometimes rooting at lower nodes, pubescent with long spreading hairs, basal leaves pinnately 3-5-foliate, long-hairy, stem leaves alternate, similar to basal leaves; 5 sepals yellowish, often tinged purple, reflexed, deciduous, 5 deep yellow petals, as long as sepals; achenes in elongate cluster, hairless with long awl-shaped beak. Moist meadows, beaches and along streams from southeastern Alaska west to Kodiak Island. Flowers July-August. Includes ***R. pacificus***.

Ranunculus nivalis (Snow Buttercup) Stems erect or ascending 10-30 cm tall, not rooting at nodes, hairy to hairless, basal leaves 3-lobed, lobes entire or again lobed, 2-3 alternate stem leaves, bracts 3-5-lobed, stalkless; flower pedicels solitary, brown-hairy, rarely hairless, 5 sepals, brown-hairy, deciduous, 5 yellow petals, longer than sepals; achenes in elongate cluster, hairless. Moist meadows, bogs and stream banks in arctic and alpine tundra most of Alaska. Flowers June-July.

Ranunculus orthorhynchus (Straight-beak Buttercup) Stems reclining to erect 40-80 cm tall, not rooting at nodes, densely to sparsely long-hairy, basal leaves pinnately 3-7-foliate, stem leaves alternate, bracts deeply lobed to simple, stalkless or short-stalked; sepals greenish, tinged with purple, reflexed, hairless, deciduous, 5 yellow petals; achenes in hemispheric clusters with straight, slender beak. Moist open sites in southeastern Alaska. Flowers July-August.

Ranunculus pygmaeus (Pygmy Buttercup) 10-15 cm tall, stems erect or ascending, pubescent with white hairs, not rooting at nodes; basal leaves simple or compound, blades 3-parted or 3-foliate, central lobe entire or 3-lobed, lateral lobes 2-4-lobed, stem leaves alternate, bracts 3-lobed and stalkless, pedicels pubescent; 5 sepals greenish often tinged purple, deciduous, 5 yellow petals; achenes in elongate cluster, with straight beak hooked at apex. Moist meadows and beaches in arctic and alpine tundra in much of Alaska. Flowers June-July.

Ranunculus repens (Creeping Buttercup) Common adventive weed from Europe naturalized in open sites and on roadsides; large showy yellow flowers, 3-foliate, margins fringed with hair, each leaf part deeply 3-lobed, toothed and white-spotted. Much of Alaska. Flowers July-August. This introduced species included because it is commonly encountered along trails.

Ranunculus verecundus (Modest Buttercup) Stems ascending to erect 5-10 cm tall, not rooting at nodes, basal leaves kidney-shaped to almost round, 3-lobed, lobes again lobed, stem leaves alternate; flower pedicels hairless, 5 yellowish (tinged purple) sepals, spreading, pubescent, deciduous; achenes in an elongate cluster, beak curved. Alpine areas in coastal southeastern Alaska from Russell Fiord to Egg Island.

Thalictrum occidentale   **Western Meadowrue**

Perennial herb with stems 30-100 cm tall from rhizome. Young stems purplish, mostly stem leaves, bluish-green, 3-4 times ternate. Flowers with male and female flowers on separate plants, female flowers with tiny greenish-white sepals and no petals, male flowers with hanging stamens and purple anthers, anthers 1.5-4 mm long, sepals greenish or purplish. Fruit achenes 4-10 mm long, stipe 0.4-1.2 mm long.

Meadows and open forests. Southeastern Alaska. Occasional. Flowers June-July.

Thalictrum alpinum (Alpine Meadowrue) Smaller than **T. occidentale**, 10-25 cm tall, leaves basal but stem can have single leaf near base; biternate, elongated racemes, flowers with both stamens and pistils, sepals purplish tinged, staminal filaments linear or slightly expanded apically, anthers 1.5-3 mm long; achenes almost stalkless. Found occasionally in bogs, on rock outcrops and in arctic and alpine tundra in most of Alaska. Flowers June-July.

Thalictrum sparsiflorum   (Few-flowered Meadowrue) Stem leaves 2-3 times ternate with small panicles; flowers with both stamens and pistils, sepals whitish or greenish, often purple tinged, staminal filaments broadened upwards, anthers 0.6-0.8 mm long; achenes long-stalked. Meadows, thickets, rock headlands and woods in much of Alaska; locally common. Flowers June-August.

148

ROSACEAE
ROSE FAMILY

KEY TO GENERA OF HERBACEOUS *ROSACEAE:*

1a. Petals lacking; many small flowers borne in dense spikes; leaves pinnately compound.   *Sanguisorba*

1b. Petals present; many flowers not at once small and not borne in dense spike, or if so, then leaves ternately dissected. **(2)**

2a. Leaves bi- or triternately dissected into linear segments; petals white. **(3)**

2b. Leaves various, but not as above; petals white, yellow or pink. **(4)**

   3a. Flowers in cymes, stems erect from taproot.   *Chamaerhodos*

   3b. Flowers raceme-like; stems prostrate from rhizomes or stolons.   *Luetkea*

4a. Leaves ternate to triternate, segments broad; plants mostly 1-2 m tall, dioecious; flowers small (petals about 1 mm long), numerous, borne in panicles to 5 cm long.
                                                                   Aruncus

4b. Leaves various, but not as above; plants mostly less than 0.5 m tall, mostly perfect; flowers larger, not in panicles. **(5)**

5a. Flowers solitary on scape-like peduncles; leaves simple, entire or with round or blunt teeth; fruit of plume-like achenes; sepals and petals 8-10 each.   *Dryas*

5b. Flowers usually more than 1; leaves compound or lobed, rarely simple; fruit plume-like achenes; sepals and petals usually 5 each. **(6)**

  6a. Leaflets tridentate apically (entire along side); stamens 5; plants prostrate or mat-forming.   *Sibbaldia*

6b. Leaflets variously toothed or lobed, but not regularly tridentate apically, or leaves simple; stamens 10 or more; plants usually not prostrate or mat-forming. **(7)**

7a. Calyx lacking bractlets between sepals; fruit an aggregate.   *Rubus*

7b. Calyx with bractlets alternating with sepals; fruit an achene. **(8)**

8a. Leaves trifoliolate; plants with well-developed stolons; flowers white; receptacle ripening into an accessory fruit.   *Fragaria*

8b. Leaves mostly with more than 3 leaflets, but if 3-foliolate or rarely simple, then lacking stolons; flowers mostly yellow (rarely white); receptacle not ripening. **(9)**

9a. Leaves palmately lobed or compound, or if pinnately lobed or compound, then not lyreshaped-pinnatifid; styles at maturity not elongate and conspicuous.
                                                                 Potentilla

9b. Leaves pinnately lobed or compound, or more usually lyreshaped-pinnatifid, terminal segment often greatly enlarged, or rarely simple; styles at maturity elongate and conspicuous.   *Geum*

KEY TO GENERA OF *ROSACEAE* TREES AND SHRUBS:

1a. Leaves compound. **(2)**

1b. Leaves simple. **(5)**

  2a. Stems usually with prickles; pistils several, separate, enclosed in hollow receptacle, which at maturity ripens into a hip.   *Rosa*

2b. Stems armed or unarmed; pistils various, but not as above; fruit not a hip. **(3)**

3a. Flowers in corymbs; ovary inferior, carpels united; fruit apple-like (a pome).
                                                                   Sorbus

3b. Flowers not in corymbs, ovary superior; fruit an achene or aggregate. **(4)**

4a. Calyx with bractlets alternating with sepals; fruit of dry achenes; stems unarmed.   *Potentilla*

4b. Calyx lacking bractlets; fruit fleshy (an aggregate); stems sometimes armed.
                                                                   Rubus

5a. Plants armed with thorns. *Crataegus*
5b. Plants unarmed, lacking thorns. **(6)**
6a. Ovary inferior (fused to calyx tube), with 2-5 united carpels. **(7)**
6b. Ovary superior, carpels 1-several, separate. **(8)**
    7a. Flowers in corymbs; fruit with 2-5 locules, each with 1-2 seeds. *Pyrus*
    7b. Flowers in racemes; fruit appearing to have 10 locules, each with single seed.
    Amelanchier
8a. Plants with prostrate branches; flowers solitary on scape-like peduncles; fruit of
    feather-like achenes; sepals and petals 8-10 each. *Dryas*
8b. Plants various, but not as above; flowers 1-several, not on scape-like peduncles;
    fruit not of feather-like achenes. **(9)**
9a. Leaves palmately lobed. **(10)**
9b. Leaves with sharp, forward pointing teeth (sometimes minute), not palmately
    lobed. **(11)**
10a. Flowers large, solitary, or in few-flowered cymes; fruit fleshy aggregate
    (raspberry). *Rubus*
10b. Flowers small, in terminal flat-topped coymbs; fruit a cluster of papery carpels.
    Physocarpus
11a. Leaves with glands on petioles; carpels 1; fruit a drupe. *Prunus*
11b. Leaves lacking glands on petioles; carpels usually 5; fruit a follicle. *Spiraea*

Amelanchier alnifolia Northern Serviceberry

Low shrub to small tree 1-5 m tall, bark dark gray to reddish. Leaves on slender petioles, hairy when young, blades nearly round with teeth toward apex. Flowers of 5 narrow lobes, densely woolly when young, in fragrant clusters, petals white. Fruit apple-like, purple or nearly black, covered with a bloom, sweet and juicy with several flattened elliptic brown seeds.

Juneau north and west to Aleutians and central Interior. Open woods, alpine and meadows. Occasional, especially around Haines and Skagway. Flowers June-August, fruit matures July-August. Fruit eaten fresh or in puddings, pies and muffins. Dried berries used like raisins. Wood hard and straight-grained and used after being fire-hardened to make arrows; also as digging sticks, spears, harpoons shafts and implement handles. When people passed away in the winter, they could not be buried until spring in cold climates where digging during winter was not possible. Amelanchiers bloom early, often as soon as the ground thaws. Consequently they were used at funeral "services", giving this plant its common name. Highly variable species. Also known as *A. florida*.

Aruncus dioicus   Goatsbeard

Perennial woody shrub 1-2 m tall, root thick, plant erect or ascending. Leaves large and long-petiolated, smaller with fewer segments above, segments with sharp forward pointing teeth. Flowers numerous in large panicles on pubescent branches, staminate and pistillate flowers on separate plants, petals white. Follicles erect with divergent tip.

Southeastern Alaska west to Attu and Shumagin islands. Forest edges, meadows, stream banks and roadsides. Common. Flowers July-early August. Root used for curing diseases of the blood, as cough medicine and treating smallpox. Also known as *A. sylvester*.

Crataegus douglasii   Black Hawthorn

Deciduous shrub or small tree to 10 m tall, young twigs reddish brown, bark gray or brown, thin, smooth or slightly fissured. Leaves thick, leathery, longer than broad, elliptic to oblong, rounded at both ends, coarsely toothed above middle. Flowers in erect clusters with several stinky flowers with 5 pointed persistent sepals, 5 oblong white petals, about 20 stamens, pistil with inferior ovary, 5 styles. Fruit apple-like, purple and covered with bloom, sweet and juicy with few dark brown seeds.

Reported from Wiehl Mt near Ketchikan and from Juneau. Forest edges, bogs, low to mid elevation. Uncommon. Flowers June-July; fruits August-September. Bark used to treat venereal disease, thin the blood and used in steam baths. Fruit edible fresh or in puddings and pies; dried berries used like raisins; commonly eaten by birds.

Dryas drummondii   Yellow Dryas

Low prostrate evergreen shrub with white-hairy petioles often with stalked reddish glands, stem from long woody base, often with small basal leaflets. Leaves basal, margins with coarse, sharp, forward pointing teeth, strongly nerved. Flowers never fully open, nodding at peak flowering, petals forming tube or funnel, sepals densely covered with long red sticky hairs, petals pale yellow. Achene with long white plumes.

Northern southeastern, south-central and northeastern Alaska. Arctic and alpine tundra to lowland areas especially gravel bars in rivers. Common. Flowers June-August. Only distinct species of *Dryas*; others form hybrid swarms.

Dryas integrifolia   Entire-leaf Mountain-avens

Matted evergreen shrub with prostrate stems. Leaves needle-like or flat, short and small, entire or toothed only near base, wrinkled, dark green and mostly lustrous above, densely white-hairy below, with strongly rolled margins. Flowers scapes woolly-haired, often with long dark glands, flower solitary, sepals narrow, acute, petals creamy white. Fruit head-like with long styles forming whitish feathery twisted plumes.

Glacier Bay and Lynn Canal area and interior Alaska. Alpine and arctic tundra, gravel bars and rocky slopes. Occasional. Flowers May-August; fruits mature June-August.

Dryas octopetala   White Mountain-avens

Decumbent or matted with woody base. Leaves oblong to ovate, rounded teeth along entire margin, hairless above, white soft matted hairs below, mid vein more or less with brown scales and long white hairs. Flowers solitary on white soft matted glandular hairy stems, calyx with long, dark glands, petals longer than sepals, white or rarely yellow. Nutlets with elongated, feathery styles.

Northern southeastern Alaska throughout most of Alaska except southwestern portion. Arctic and alpine tundra, woodlands. Common. Flowers July-August.

Fragaria chiloensis   Beach Strawberry

Perennial coastal herb 2-25 cm long from stout rhizome and long brown stolons, petioles often reddish tinged. Leaves mostly basal, 3-foliate, thick, leathery, leaflets stalked and covered with long soft woolly hairs, veins conspicuously elevated. Petals 5, obovate to round, 5 sepals alternating with 5 sepal-like small bracts, petals white. Achenes with soft hairs, borne on fleshy receptacle which ripens to form an edible fruit.

Southeastern Alaska west to Aleutians. Meadows, gravelly beaches, rock crevices and woods. Occasional. Flowers late May-July. Hybrids with cultivated strawberries occur. Fruit palatable; makes excellent jam. Strawberry leaves used as an anti-diarrhea medicine.

Geum calthifolium   Caltha-leaf Avens

Hairy perennial herb to 25 cm tall with stem from stout, dark brown woody base. Basal leaves long-stalked, large, round to kidney-shaped with rounded teeth in margins, hairy on both sides with short yellow hairs; stem leaves paired and reduced upwards. Petals twice as long as calyx, heart-shaped, often with 1-3 stalkless bracts below flower, petals yellow. Achene with rough hairs or bristles.

Southeastern Alaska west to Aleutians. Wet meadows, bogs and ridges at low elevation. Common. Avens species claimed to ward off evil spirits. Flowers June-August.

Geum macrophyllum   Large-leaf Avens

Hairy perennial herb to 1 m tall with long soft white hairs on stems and petioles. Basal leaves large on long petioles, interruptedly lyreshaped-pinnate, terminal leaflet largest, toothed and conspicuously lobed, hairy along veins below. Styles elongate with s-shaped bend near tip, often glandular below bend, petals longer than reflexed sepals, calyx lobes triangular, petals yellow. Sepals and styles reflexed in fruit, nutlets hairy.

Southeastern Alaska north to Brooks Range. Wet meadows, roadsides, beaches and woods. Common. Flowers June-July. Bitter tonic made from whole plant used to increase appetite after illness, as gargle for sore throat, or to get rid of bad breath.

Luetkea pectinata   Partridgefoot

Prostrate matted perennial herb 3-15 cm tall. Leaves fan-shaped, hairless with several linear, somewhat acute lobes. Flower peduncles erect with smaller leaves, branches of inflorescence hairy, calyx lobes triangular, petals obovate, white. Fruit several-seeded follicle on stipe.

Southeastern Alaska west to Aleutians. Alpine and subalpine meadows or stream gravels. Common. Flowers July-September.

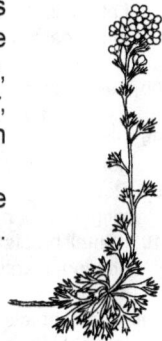

Physocarpus capitatus　　Pacific Ninebark

Deciduous shrub to 4 m tall with shedding bark. Leaves alternate with narrow paired stipules shredding early, blades ovate to heart-shaped, palmately 3-5 lobed about halfway to midrib, double toothed, angled twigs, bark peeling in long strips exposing orange-brown inner bark. Flowers in terminal clusters, flattened, flowers in cup shaped base, 5 long pointed sepals, flower color white to greenish. Fruit pod-like, 3-5 and opening on two lines, persistent in winter.

Southern southeastern Alaska. Forests, banks of streams. Uncommon and local. Flowers July-August; fruits July-August. Bark has laxative effect.

KEY TO SPECIES OF *POTENTILLA:*

1a. Flowers reddish purple to dark red or maroon; petals less than half as long as sepals. *P. palustris*

1b. Flowers white, cream, or yellow, not at all red or purple. **(2)**

2a. Plants strongly stoloniferous; flowers solitary on naked scapes. *P. anserina*

2b. Plants lacking stolons; flowers one to several on leafy stems. **(3)**

3a. Leaves green above and below, or white-tomentose only on lower surface. **(4)**

3b. Leaves with silky soft hairs on both surfaces. *P. hippiana*

4a. Leaves palmately compound, with 5 or more leaflets (at least some). *P. gracilis*

4b. Leaves 3-foliolate. **(5)**

5a. Plants annual or biennial, from taproots; leaves mostly on stem, green on both surfaces, never white-hairy or with dense woolly hairs below; stems and lower leaf surfaces with stiff hairs. *P. norvegica*

5b. Plants perennial from woody base or rhizomes; leaves mostly basal, variously pubescent, but most species with white, densely woolly hairs on lower surface. **(6)**

6a. Leaflets without dense whitish or grayish woolly hairs on either side, variously pubescent to hairless on both surfaces. **(7)**

6b. Leaflets with dense, woolly hairs on one or both surfaces, upper surface mostly sharply contrasting in color with lower. **(9)**

　7a. Plants mostly 1.5-4 cm tall; leaves (0.5)1.5-2 cm long, terminal leaflet 2.5-7 mm long and 2.5-5 mm broad; sepals 1.8-2.5 mm long. *P. elegans*

　7b. Plants mostly 3-20 cm tall; leaves 2-10 cm long, terminal leaflet 7-22 mm long or more and 5-17 mm broad; sepals 3.5-6.5 mm long. **(8)**

8a. Lateral leaflets usually deeply incised into 2 narrowly oblong lobes, and terminal into 3 narrow lobes, usually covered with whitish bloom below, with rolled margins. *P. biflora*

8b. Lateral leaflets and terminal ones 3-11-toothed, not at all with whitish bloom below, usually margins not rolled. *P. hyparctica*

9a. Petioles and at least lower portions of stems cobwebby-densely woolly, hairs not soft and long. *P. nivea*

9b. Petioles and stems with soft or long hairs (sometimes densely woolly beneath longer outer hairs). **(10)**

10a. Small bracts of calyx mostly 1.5-2 times longer than broad, ovate to lance-shaped; plants mostly with long, soft somewhat wavy hairs throughout, usually coastal. *P. villosa*

10b. Small bracts of calyx commonly 2-4 times longer than broad, narrowly oblong to lance-shaped; plants variable in pubescence, usually not long-villous throughout, mostly interior, rarely coastal. *P. hookeriana*

154

Potentilla anserina   Common Silverweed

Silky-haired perennial herb with long, strawberry-like stolons. Leaves basal to 40 cm long, pinnately compound with 5-17 main leaflets interspersed with smaller ones, with coarse, sharp, forward pointing teeth, green and hairless above, white hairy beneath. Flowers on sparsely white-haired scape to 2 cm tall, petals oval to ovate, sepals ovate, erect and enlarging in fruit, petals yellow. Achenes with shallow furrow on back.

Along coast and river systems of Alaska. Tidal meadows, beaches, intertidal, sometimes covered by tide water. Common. Flowers June-August. Roots dug in late fall or early spring and eaten raw, boiled, steamed, or roasted; tastes like parsnip. Includes *P. egedii*. Also known as *P. pacifica.*

Potentilla palustris   Marsh Cinquefoil

Perennial herb 10-100 cm long from creeping, somewhat woody rootstock, stems prostrate to ascending, purplish. Leaves pinnate with 5-7 leaflets, oblong, lance-shaped to elliptic, margins coarsely toothed, dark green above, pale and hairy below. Flowers with fetid odor, showy, few to several, petals about half as long as sepals, brownish reddish purple to intense purple. Numerous achenes 1-1.5 mm long.

Throughout Alaska. Wet meadows, along streams and in shallow fresh water. Common. Flowers July-August. Dried leaves make medicinal tea. Our only cinquefoil with red rather than yellow flowers.

Potentilla villosa   Villous Cinquefoil

Perennial herb to 30 cm tall, stems stout from short branched woody base, densely covered with white woolly hairs. Leaves leathery, 3-foliate, with coarse, sharp, forward pointing teeth, dark grayish-green and hairy above, strongly ribbed below with greenish-white hairs. Few large showy flowers with shallow notch at tip of yellow petals. Achenes several to many, 1-1.5 mm long.

Southeastern Alaska west to Aleutians, less common in central Interior Alaska. Beaches, cliffs, rock outcrops up to alpine and meadows. Common. Flowers June-August.

Potentilla biflora (Two-flower Cinquefoil) Perennial herb 4-15 cm tall, basal leaves pinnately 3-foliate, lateral leaflets commonly deeply cleft, terminal with 3 narrowly oblong lobes, oval and sparsely hairy to hairless above, less hairy or hairless below, rolled margin; one to several showy flowers, sepals ovate-lance-shaped, small lance-shaped to oblong bracts, yellow petals with shallow notch at apex; achenes numerous. Hillsides in tundra most of Alaska north to Brooks Range.

Potentilla elegans (Elegant Cinquefoil) Dwarf perennial herb 1.5-4 cm tall from woody base; leaves palmately 3-foliate, obovate, apically toothed, lateral leaflets mostly 5-toothed; flowers solitary, inconspicuous, petals pale yellow to white. Alpine regions, often on rocky outcrops from widely separated areas throughout Alaska including Mt. Roberts near Juneau.

Potentilla gracilis (Slender Cinquefoil) 5-100 cm tall, non-glandular, basal leaves palmately compound with 5-9 leaflets, terminal one coarsely toothed near base; flowers showy, several to many, sepals ovate-lance-shaped, small lance-shaped bracts, petals yellow, shallowly notched at apex; achenes numerous. Woodlands, thickets and meadows in coastal southern Alaska and in major drainages in eastern Alaska. Flowers July-August.

Potentilla hippiana (Woolly Cinquefoil) 30-50 cm tall, with long wavy silky hairs, basal leaves pinnately 5-7-foliate, leaflets white or grayish silky, coarsely toothed; several to many showy flowers, sepals and small lance-shaped bracts shorter than sepals, petals yellow, shallowly notched at apex; achenes numerous. Stikine River and one site in Alaska Range. Rare.

Potentilla hookeriana (Hooker Cinquefoil) 8-25(50) cm tall, from woody base, stems and petioles with long spreading hairs, basal leaves palmately compound, with 3(5) leaflets, terminal one elliptic to oblong or obovate, coarsely toothed to near base, sinus often extending to near midrib, green with appressed, stiff, short hairs above, pale and densely woolly hairs below; flowers few to several, small but showy, petals yellow, with shallow notch at apex; achenes several. Gravelly soils and rocky outcrops throughout Alaska. Flowers June-July.

156

Potentilla hyparctica (Arctic Cinquefoil) Perennial 3-20 cm tall non-glandular herb, stems and petioles with long spreading hairs, basal leaves palmately 3-foliate, terminal leaflet 3-11-toothed near base, silky haired especially along veins; 1-few showy flowers, sepals lance-shaped, small lance-shaped to oblong bracts, petals yellow, shallow notch at apex; several achenes. Coastal beaches, arctic and alpine tundra, most of Alaska except central valleys. Aleutian plants and high elevation specimens dwarfed. Flowers July-August.

Potentilla nivea (Snow Cinquefoil) Perennial non-glandular herb 7-30 cm tall from woody base, stems and petioles cobwebby-densely woolly; basal leaves palmately compound with 3(5) leaflets; flowers 1-several, showy, sepals lance-shaped, petals yellow. Meadows, open slopes and rock outcrops mostly in Brooks Range, but also in Alaska and St. Elias Ranges and on Mt. Roberts near Juneau. Closely related to *P. hookeriana* and *P. uniflora*.

Potentilla norvegica (Rough Cinquefoil) 10-100 cm tall, stem and petiole sparsely stiff-hairy, leaves mostly on stems (not basal), palmately compound with 3(5) leaflets, terminal one coarsely toothed near base or entire and wedge-shaped in lower part, sparsely stiff-hairy, especially along veins; several to many inconspicuous flowers, sepals ovate-lance-shaped, small oblong, elliptic or lance-shaped bracts, petals yellow or whitish, shallow notch at apex; achenes numerous. Disturbed soils along roads, old fields, stream banks and meadows in much of Alaska south of Brooks Range except western and southwestern parts. Flowers July-August. May be both native and introduced.

Malus fusca Oregon Crab Apple

Small deciduous tree 2-12 m tall usually with several trunks or thicket-forming shrub, bark gray, smooth to slightly scaly. Leaves ovate to elliptic on slender petioles, 3 irregular lobes. Flowers in clusters on slender stalks with 5 pointed hairy sepals and 5 rounded petals, petals white or pink. Fruit oblong like a small apple, yellow or red, with sour taste.

Southeastern Alaska, Prince William Sound to Kenai Peninsula. Beach forests, bog fringes and along streams. Occasional. Flowers in June; fruits mature August-October. Fruit sour but edible, preferred food for deer and moose. Hard wood good for bows, wedges, digging sticks and halibut hooks. Also known as *Pyrus fusca*.

Rosa nutkana   Nootka Rose

Spiny deciduous shrub to 3 m tall. Leaflets mostly 5-7, paired except for terminal leaflet, stalkless, elliptic or ovate, rounded at both ends, edges mostly doubly toothed, mostly hairy along veins. Flowers mostly one at end of short lateral twig on stout erect stalk, 5 large narrow leaf-like sepals, 5 pink to rose petals. Fruit berry-like, rounded red or purplish hip containing several to many hairy, shiny, brown seeds that persist in winter.

Southeastern Alaska west to Aleutians. Woods, meadows, forming thickets along beaches. Common. Flowers June-August; fruits August. Young tender shoots eaten in spring; hips make excellent jam, tea and wine; boiled root or bark tea good for sore throat or as an eye wash. Hybrids with **R. acicularis** where ranges overlap. **R. rugosa** (Sitka Rose) was introduced in Alaska from Asia.

KEY TO SPECIES OF *RUBUS:*

1a. Leaves simple (at least some), broadly and shallowly to deeply 3-5-lobed. **(2)**
1b. Leaves compound, with 3-5-leaflets. **(4)**

    2a. Plants distinctly woody, with large maple-like leaves, lobes acute or acuminate, usually over 40 cm tall. .... *R. parviflorus*
    2b. Plants herbaceous, or only semi-woody, leaves with broadly rounded lobes; stems trailing, usually less than 40 cm tall. **(3)**

        3a. Flowers white, imperfect, plants dioecious; fruit yellowish when ripe, flavor like that of baked apple. .... *R. chamaemorus*
        3b. Flowers pink, perfect; fruit reddish or purplish at maturity, not flavored like baked apple. .... *R. arcticus*

4a. Stems erect or arching, woody, seldom less than 50 cm tall. **(5)**
4b. Stems low, often trailing, herbaceous or only slightly woody, seldom as much as 50 cm tall. **(6)**

    5a. Petals red or pink, 15-25 cm long; fruit yellowish to red, mostly 12-20 mm broad, sour and soapy; stems unarmed or armed mostly near base. .... *R. spectabilis*
    5b. Petals white (rarely pink), 6-10 cm long; fruit red to reddish purple or black, often less than 12 mm broad; stems armed with spines or prickles. .... *R. leucodermis*

6a. Petals white, mostly 6-9 mm long; upright stems less than 5 cm tall. ... *R. pedatus*
6b. Petals pink, mostly 10-15 mm long or longer; plants 3-30 cm tall. ... *R. arcticus*

Rubus arcticus   Nagoon Berry

Herbaceous or somewhat woody unarmed perennial to 10 cm tall, more or less covered with short hairs. Leaves palmately compound or 3-lobed, lobes rounded heart-shaped to round. Flowers showy, usually solitary (rarely 2), sepals and petals long and narrow, petals pink to reddish pink. Fruit red to purplish drupelets, several to many.

Most of Alaska except northwestern portion. Meadows and bogs, forests and beach fringe, low elevation. Common. Flowers May-early July. Berries palatable, richly flavored, excellent for jam and flavoring liquor. Includes *R. stellatus*; also known as *R. arcticus* ssp *stellatus*.

Rubus chamaemorus   Cloudberry

Perennial unarmed herb 2-20 cm tall, flowering stem erect from thin, creeping woody rootstock. Leaves leathery, obscurely palmately lobed and veined, margins sharply toothed. Flowers solitary, showy, 5 broad petals, 5 hairy sepals, male and female flowers on different plants, petals white. Fruit drupelets hard and red-tinged when immature, when loosening from sepals, yellow and juicy.

Throughout Alaska. Bogs, meadows and alpine. Common. Flowers May-June. Fruit excellent; mixed with fat to make Eskimo ice cream, high in vitamin C, ripened fruit tastes a little like apricots or baked apple. Cloudberry plants are either male or female.

Rubus parviflorus   Thimbleberry

Deciduous erect shrub 40-150 cm tall. Leaves simple, maple-like with paired lance-shaped stipules and slender petioles, blades palmately lobed with sharply double toothed margins, soft hairy. Flowers in flat-topped terminal clusters, mostly 3-7 with 5 spreading, narrow hairy greenish sepals, 5 obovate white petals, numerous stamens and pistils. Fruit thimble-like, large, half-round seedy berry.

Southeastern Alaska. Edge of forest, rocky areas and forests along beaches. Locally common, uncommon on Admiralty, Baranof and Chichagof islands in southeastern Alaska. Flowers June-July; fruit ripens August-September. Berries eaten raw or dried. Young sprouts peeled and eaten raw. Large leaves used to line cooking pots and cover baskets of berries.

Rubus pedatus   Five-leaf Bramble

Creeping, perennial herb from slender trailing vine rooting at nodes. Leaves alternate, palmately compound with 5 leaf divisions, stalkless, oval, irregularly toothed. Flowers small, 1-2 with petals and sepals about equal, petals white. Fruit 1-6 red drupelets.

Southeastern Alaska west to Shumagin Islands. Moist, mossy forests, adjacent meadows and bogs. Common. Flowers June-July; fruits July-September. Fruit juicy and delicious but occurs in small quantity; makes excellent jam.

Rubus spectabilis   Salmonberry

Hairy and somewhat prickly woody shrub to 4 m tall. Leaf stipules linear, leaves pinnately 3 or rarely 5-foliate. Flowers showy on short, leafy shoots, solitary, staminal filaments somewhat flattened, petals reddish-purple or orange. Fruit raspberry-like, orange or red.

Southeastern Alaska west to Aleutians. Wet areas, forest openings, along streams, recent avalanche tracks, forms thickets with Devil's Club in subalpine regions and recent clearcuts. Common. Flowers April-July; fruit matures in early July south, not until August in north of range and at high elevation. Berries juicy and used for jams and jelly. Flower nectar sweet and tasty in season. Salmonberry forms extensive clones some having tastier berries than others. Sprouts in spring peeled and eaten raw or steamed. Stems used as pipe stems, for arrow shafts and pegs. Berries eaten by bears.

Rubus leucodermis (Western Black Raspberry) Erect shrub 1-2 m tall, stems and petioles armed with hooked spines; leaves pinnately 3-5-foliate, green and hairless or sparsely hairy above, white hairy below; petals white with slender staminal filaments; fruit raspberry-like, several to many, reddish purple to black, good flavored. Forest edge and thickets along coast and on islands of southeastern and south-central Alaska. Uncommon. Flowers June-early July. Young shoots peeled and eaten raw or cooked.

Sanguisorba canadensis   Sitka Burnet

Perennial hairless herb 25-100 cm tall, stem from stout woody base. Basal leaves pinnate, egg-shaped to oblong, leaflets with sharp teeth in margins; 1-3 small stem leaves. Flowers in spikes, 4 stamens, filaments much longer than sepals, flattened and expanded upwards, petals oval, greenish white. Fruit an achene, hypanthium hairless, narrowly winged.

Southeastern Alaska west to Aleutians and north to 65th parallel. Bogs, snow beds, meadows and along streams. Common. Flowers July-August. Young leaves good in salads; stops diarrhea, coagulates blood, promotes perspiration and used to cure gout. Also known as *S. stipulata* and *S. sitchensis*.

Sanguisorba menziesii   Menzies Burnet

Rhizomatous perennial herb 30-50 cm tall. Basal leaves ovate to oblong, coarsely and sharply toothed leaflets; stem leaves small. Flower spikes shorter,1.5-7 cm long, than those of *S. canadensis*, stamens 4, filaments flattened and broadened upwards, usually protruding, about twice as long as sepals, petals lacking, sepals reddish purplish. Hypanthium hairy, winged in fruit.

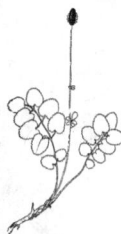

Southern southeastern Alaska; Cook Inlet area. Moist places, meadows and bogs. Less common than *S. stipulata*. Flowers July-August. *S. menziesii* morphologically intermediate between *S. officinalis* and *S. canadensis*; may have originated from hybridization of these two species.

Sanguisorba officinalis   Common Burnet

Perennial herb 20-80 cm tall from stout rhizome. Basal leaves 10-30 cm long with 7-15 ovate to oblong or lance-shaped, coarsely toothed leaflets; 1-2 small stem leaves. Flower spikes 1-3 cm long, sepals 2-3 mm long, stamens 4, almost equal to sepals or somewhat longer, filaments linear, hypanthium hairy, flower color reddish purple. Hypanthium winged.

Southeastern Alaska, Seward Peninsula and Point Hope eastward through central interior Alaska. Woods, bogs and tundra low to mid elevation. Flowers July. Also known as *S. microcephala*.

Sibbaldia procumbens   Sibbaldia

Low, mat-forming perennial herb from woody base. Basal leaves wedge-shaped, ternate, 3-toothed at apex, covered on both sides with stiff, straight, flat lying hairs, stem short. Flowers crowded, petals minute, linear-oblong, borne in leafy bracted cymes, pale yellow. Achenes on stipe.

Southeastern Alaska west to Aleutians north to Denali Park. Alpine meadows, snow beds and open woods. Occasional. Flowers late June-August. Similar to cinquefoil; except Sibbaldia has tiny petals and 5 or 10 stamens.

Sorbus sitchensis   Sitka Mountain Ash

Deciduous shrub or small tree 1-4 m tall. Leaves pinnate with paired narrow rusty-haired stipules, leaflets usually 9 or 11, elliptic or oblong, edges coarsely and sharply toothed above middle, twigs with odor and bitter cherry taste, bark smooth gray. Flowers in rounded terminal clusters with 15-60 flowers on rusty haired stalks, small fragrant flowers with 5 broadly triangular sepals, 5 rounded white petals. Fruit several to many, small apple-like, red becoming orange and purple with few elliptic brown seeds.

Southeastern Alaska west to Kodiak Island and Alaska Peninsula. Open woods and thickets to alpine meadows. Occasional. Flowers June-August; fruits August-September. Fruit edible but bitter.

Sorbus scopulina (Western Mountain-ash) Grayish red or yellowish bark, winter buds glutinous and glossy, twigs white-hairy; leaflets 7-13, branches of inflorescence sparsely to densely pubescent with white hairs; petals white to cream; fruit scarlet to orange, drying purplish. Woods, thickets and alpine tundra from southeastern Alaska west to Alaskan Peninsula and north to 65th parallel. Fruit edible and makes a tart jam. Flowers June-July. *S. aucuparia* (Mountain Ash) is the introduced Rowen Tree found in towns and settlements throughout southeastern Alaska.

Spiraea stevenii   Alaska Spiraea

Small, much branched deciduous shrub to 1 m
tall with hairy reddish branchlets. Leaves elliptic to
ovate on short pedicels rounded at both ends, edges
sharply toothed nearly to base. Flowers in flattened to
half round terminal clusters of many crowded flowers
with 5 triangular sepals bent down, 5 white petals,
many white stamens about twice as long as petals.
Fruit five shiny brown pod-like follicles, finely hairy,
persistent in winter.

Juneau area north and west to eastern half of
Alaska Peninsula north throughout most of Alaska
except northern coastal plain. Wet meadows, bogs and alpine.
Occasional. Flowers June-August; fruits July-September. Leaves
brewed as tea. Also known as *S. beauveriana.*

Spiraea douglasii   Douglas Spiraea

Erect deciduous shrub to 2 m tall, twigs
slender, reddish brown with soft hairs, becoming
dark brown and hairless. Leaves on short,
hairy petioles, blades elliptic to oblong, edges
sharply toothed in upper half. Flowers in terminal
clusters, cone-like, several times as long as
broad, finely hairy, many crowded short-stalked
flowers with 5 triangular sepals bent down, 5
petals, round to obovate, many pink to rose
stamens, 5 pistils, petals pink to rose. Fruit long,
shiny brown pod-like follicles, hairless or nearly
so containing two to several narrow seeds.

Dixon Entrance to Sitka and Hoonah.
Banks of streams, bogs, lakes and wet meadows.
Uncommon. Flowers June-August; fruits
September. Branches used to make hooks for
drying and smoking salmon. Fire-hardened
wood used for blades, halibut hooks and to
scrape cambium.

RUBIACEAE
MADDER FAMILY

Galium aparine   Cleavers

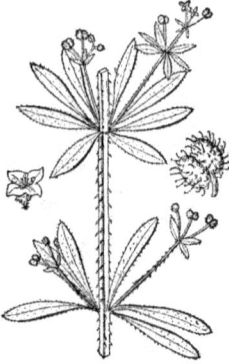

Annual herb 30-100 cm tall, readily adheres to clothing, plant emerges each year from taproots, stem sprawling, quadrangular with downward directed prickles on stem. Leaves linear to oblong in whorls of 6-8, bristle-tipped, 1-nerved, with backward facing prickles in margin. Mostly 3-5 flowers, subtended by leaf-like bracts, borne in cymes on axillary peduncles with 4 sharply pointed unattached lobes, petals white. Fruit twin seeds in spherical spiny heads.

Southeastern Alaska west to Aleutians. Beaches, open areas and meadows. Common. Flowers July-August. The young greens edible. This and other *Galium* species can be rubbed on hands to remove pitch. Dried plants used as tinder for lighting fires. The round burs are collected while greenish, roasted until brown and then ground as a coffee substitute. *Rubiaceae* is also the Coffee Family.

Galium kamtschaticum   Boreal Wild-licorice

Rhizomatous, perennial herb 10-30 cm tall, stems single, erect, sparsely hairy or hairless. Leaves oval to broadly elliptic or obovate in whorls of 4, marginal hairs pointed forward, 3-veined, bristle-tipped apically. Flowers mostly 1-3(6) from uppermost whorl, corolla 4-lobed with acute lobes, petals greenish white. Fruit pubescent with hooked bristles.

Southeastern Alaska west to Aleutians. Moist places, especially alluvial, talus slopes low to subalpine elevation. Uncommon. Flowers July-August.

Galium trifidum   Small Bedstraw

Rhizomatous, perennial herb 5-70 cm tall, stems sprawling, minute bristles turned backward along stem angles. Leaves oblong to linear or elliptic, 1-veined, in whorls of 4(5-6), with backward turned hairs along margin, 1-veined, rounded apically. Flowers 1-3 borne in axillary or terminal cymes, corolla with obtuse lobes, 4-lobes, petals whitish. Fruit about 1 mm long, smooth nutlet.

Most of Alaska. Moist open areas, beaches. Common. Flowers late June to August. Plant juice or a strong tea is used for bathing slow-healing burns, open sores, inflamed stretch marks or any festering skin conditions. Also used in hot packs for relieving aches and pains and to make an infusion taken as a remedy for urinary tract problems.

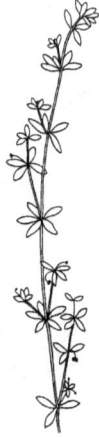

Galium triflorum   Sweet-scented Bedstraw

Dark-green perennial herb 20-100 cm long from rhizome, stems sprawling with backward facing bristles to smooth on margins. Leaves broad, elliptic, in whorls of 5-6 (4 on some branches), marginal hairs pointing forward, 1-veined, tipped with sharp rigged point. Flowers mostly 3 (rarely more) in axillary cymes, corolla 4-lobed, petals whitish. Fruit covered with soft, short hairs and hooked prickles.

Southeastern Alaska west to Alaska Peninsula. Open areas, moist forests low to mid elevation. Common. Flowers June-August. Releases sweet smelling coumarins when crushed. Plant used as hair rinse; dried flowers as perfume. Rich in vitamin C, important for explorers of old to prevent scurvy. Compresses prepared to stop bleeding and soothe aching muscles. Dried leaves made into wash to treat eczema and psoriasis. Tea used as weight loss aid and to lower blood pressure. Small seeds roasted as coffee substitute. Caution: People with diabetic tendencies or poor circulation should avoid this plant. Extensive use causes mouth irritation.

SALICACEAE
WILLOW FAMILY

Populus balsamifera   Black Cottonwood

Large deciduous tree to 30 m tall and over 100 cm in diameter with narrow pointed crown; in age larger and developing a massive trunk and small flat-topped crown, twigs red brown, hairy with orange lenticels becoming dark gray, bark smooth and greenish when young becoming grayish and deeply furrowed in age. Leaves with long finely hairy petioles channeled above, leaf blades broadly ovate with many small rounded teeth, shiny dark green above, whitish often with rusty specks below. Catkins drooping with many small flowers. Fruit pubescent capsules with 3 (rarely 2 or 4) valves.

Southeastern (primarily mainland) to south-central Alaska and Kodiak Island. Along streams, gravelly flats, open forests and recently deglaciated areas. Occasional. Flowers in May before leaves emerge; fruits June-July. Cambium scraped off and eaten fresh or sun-dried; sweet but ferments rapidly and cannot be stored (like Western Hemlock cambium) for winter use. Bark used for buckets to carry and store food. Cottonwood ashes used for soap. Balm made from winter buds and used for sores, rash, frostbite or to relieve congestion. Also known as *P. trichocarpa. P. b. trichocarpa* and *P. b. balsamifera* hybridize where ranges overlap near Lynn Canal and Cook Inlet.

Populus tremuloides (Quaking Aspen)
Slender, small to medium tree mostly less than
10 m tall with smooth, white or greenish bark with
laterally flattened petioles and leaves that tremble
in slightest breeze. Woods and tundra in much
of continental Alaska south of Brooks Range
including north of Haines; although common
throughout much of Alaska, this species is barely
in our range.

KEY TO SPECIES OF *SALIX:*

1a. Plants dwarf or prostrate trailing shrubs, under 20 cm tall. **(2)**
1b. Plants erect shrubs exceeding 20 cm tall, or trees. **(11)**

    2a. Pistillate and staminate flowers with 2 nectaries, one on either side of
        stipe; leaf veins form a prominent network, leaves pale beneath; cat-
        kins borne on prominent, almost terminal, flower-bearing branchlets.
                                                       S. reticulata
    2b. Pistillate flowers with 1 nectary between stipe and catkin axis; catkins borne on
        lateral, flower-bearing branchlets. **(3)**
    3a. Pistils pubescent, sometimes only on beak (see also *S. stolonifera*). **(4)**
    3b. Pistils hairless. **(8)**
    4a. Leaves green or pale green beneath, not with white bloom.   *S. polaris*
    4b. Leaves with white bloom beneath. **(5)**
    5a. Leaves 0.9-1.5 cm long, margins prominently ciliate; catkins spherical.
                                                       S. ovalifolia
    5b. Leaves longer than 1.5 cm, margins not ciliate (except in some *S. arctica*);
        catkins cylindrical. **(6)**
    6a. Styles 0.1-0.5 mm long.   *S. fuscescens*
    6b. Styles longer than 0.5 mm long. **(7)**
    7a. Pistils sparsely or densely pubescent with crinkly, refractive hairs; nectaries
        equal to or longer than stipes; branchlets and leaves various. **(8)**
    7b. Pistils densely pubescent; leaves dark green and usually glossy
        above, wedge-shaped to rounded at base; branchlets trailing to erect.
                                                       S. arctica
    8a. Leaves green (not having white bloom beneath).   *S. rotundifolia*
    8b. Leaves with white bloom beneath. **(9)**
    9a. Branchlets usually densely woolly; leaves lemon green, leathery, obovate to
        narrowly obovate and tapering to short petiole; petioles to 0.3 mm long; bracts
        tawny, apex often notched; pistils commonly brick red.
                                                 S. setchelliana
    9b. Branchlets hairless to sparsely pubescent; leaves thin, elliptic to almost circular;
        petioles 2-20 mm long; bracts brown to blackish; pistils reddish, purple, or
        greenish. **(10)**

    10a. Branches short and erect, sometimes trailing, often with white bloom; plants
        often rhizomatous; styles 0.8-1.6 mm long.   *S. stolonifera*
    10b. Branches long and trailing, without white bloom; styles 0.2-0.8 mm long.
                                                  S. ovalifolia

11a. Plants flowering before leaves appear. **(12)**
11b. Plants flowering as or after leaves appear. **(17)**

12a. Pistils hairless. **(13)**
12b. Pistils pubescent. **(15)**
13a. Stipules absent; branchlets brittle and with persistent, long, soft, somewhat wavy hairs at base; plants of coastal southern Alaska.
S. hookeriana
13b. Stipules present, often persistent; branchlets wiry, without long, soft, some-what wavy hairs at base; plants not of coastal Alaska. **(14)**
14a. Stipule persistent for several years, linear to ovate, apex attenuate; styles lon-ger than 1.2 mm; nectaries 2-3 times as long as stipes. *S. lanata*
14b. Stipules not persistent for more than 1 year, elliptic to broadly ovate, rounded apically, styles shorter than 1.2 mm; nectaries shorter than stipes.
S. monticola

       15a. Leaves with dense, white, interwoven hairs be-neath, bright green above; stipes to 0.4 mm long.
S. alaxensis
       15b. Leaves silky or with long, dense, somewhat wavy hairs to sparsely pubescent, smooth and hairless beneath; stipes 0.2-2 mm long. **(16)**
       16a. Branchlets velvety with soft hairs; styles 0.2-0.5 mm long; stipes 0.8-2 mm long. *S. scouleriana*
       16b. Branchlets covered with down or long, soft, somewhat wavy hairs or becoming hairless; styles 0.5-1.8 mm long; stipes 0.2-0.8 mm long. *S. planifolia*

17a. Pistils hairless. **(18)**
17b. Pistils pubescent. **(22)**

18a. Leaves green or pale beneath, without white bloom. **(19)**
18b. Leaves with white bloom beneath. **(20)**
19a. Leaves with coarse, long, somewhat wavy hairs on both sides, margins glan-dular-minutely toothed or partly entire. *S. commutata*
19b. Leaves hairless, margins round to sharply toothed. *S. myrtillifolia*
20a. Petioles glandular near leaf base; stamens 5; leaf apex acuminate to bearing a tail-like appendage. *S. lasiandra*
20b. Petioles not glandular; stamens 2; leaf apex acute to rounded. **(21)**
21a. Stipules absent; leaves pubescent beneath, at least on midrib; branchlets brittle, with long, persistent, soft, somewhat wavy hairs at base; styles red.
S. hookeriana
21b. Stipules present; leaves hairless beneath, elliptic or obovate; branchlets wiry, with dense to sparse long, soft, somewhat wavy hairs, lacking persistent hairs at base; style greenish. *S. barclayi*

22a. Leaves silky-haired beneath, margins glandular-minutely toothed to distantly so. **(23)**
22b. Leaves densely pubescent to becoming hairless beneath, not silky-haired; peti-oles yellowish. *S. glauca*
23a. Leaves narrowly lance-shaped, 5-7 times as long as wide, silky be-neath with short white or rusty-colored hairs oriented toward apex, mar-gins prominently glandular-minutely toothed; styles 0.3-0.5 mm long.
S. arbusculoides
23b. Leaves narrowly elliptic to obovate, 2.5-3 times as long as wide, appearing satiny beneath with matted, silky hairs, margins distantly and inconspicuously glandular-minutely toothed to glandular-rounded toothed; styles 0.5-0.8 mm long.
S. sitchensis

Salix arctica   Arctic Willow

Dwarf shrub to 50 cm tall, trailing or forming dense mats, branchlets glossy or sparsely pubescent with straggly hairs. Leaves grayish-green with whitish bloom, sparse hairs below, broadly oval to elliptic with entire margins, apex usually bearded with long, straight hairs. Catkins large to 10 cm long and 15 mm thick, erect on stalks, scales black to brown with long silky hairs.

Most of Alaska. Arctic and alpine tundra throughout Alaska; sea level on glacial outwash and moraines in southeastern Alaska. Common. Flowers June-July; fruits ripen July-August.

Salix barclayi   Barclay Willow

Shrub or small tree to 6 m tall, twigs yellowish-green and densely hairy when young, becoming reddish brown and hairless with blackish buds, bark smooth, gray or greenish brown, twigs often end in rounded insect-caused galls or "willow roses" composed of deformed leaves. Leaves broadly elliptic to obovate, short-pointed tip and wedge-shaped to rounded base, margins saw toothed to entire, sometimes with short reddish hairs along midrib, lower surface whitish, soon becoming hairless, usually turning black in drying. Catkins on leafy stalk appearing with leaves, scales 1 mm long, black with long hairs.

Southeastern Alaska north to southern third of Alaska. Glacial moraines, lakes and river shores, subalpine and alpine slopes, bogs and forests. Common. Flowers late May; seeds ripen in July. Difficult to distinguish from *S. hookeriana*. Forms hybrids with *S. commutata*.

Salix commutata   Undergreen Willow

Low shrub or small tree to 2 m tall, branches dark brown, twigs densely white-woolly or with long sparse soft, wavy hairs. Leaves elliptic, abruptly pointed, densely grayish woolly when young, hairless with age, stipules leaf-like, glandular margins. Catkins on leafy peduncles, densely woolly when first emerging, capsules reddish.

Northern half of coastal southeastern Alaska to Aleutians, north to central Alaskan Range. Pioneering species on glacial moraine and rocky slopes in alpine tundra, also gravelly beaches, willow thickets and bogs. Common. Flowers late May-June; seeds ripen late June-July. Hybridizes with *S. barclayi.*

Salix scouleriana   **Scouler Willow**

Shrub 2-7 m tall or tree 10-20 m tall with compact rounded crown, bark gray, smooth, thin becoming dark brown and divided into broad flat ridges. Leaves variable, mostly oblanceolate to narrowly obovate, very short-pointed at apex tapering to base, edges without teeth to sparsely wavy-toothed, dark green and nearly hairy above, whitish hairy below. Catkins stout, mostly stalkless, appearing in great abundance before leaves, scales long-hairy.

Southeastern and central Alaska, Kenai and Alaska Peninsulas. Open areas, bogs, burned areas, along streams and roadsides. Common. Flowers May; one of earliest willows to flower. One of several willows used for diamond willow carvings. Willow twigs used to clean teeth. Wood used to smoke salmon, dry meat and fish; rotten willow roots used as punk which could be ignited and carried while traveling. Also used to make spoons and other small items.

Salix sitchensis   **Sitka Willow**

Shrub or small tree 1-8 m tall with brittle twigs, in exposed places becoming low, nearly prostrate shrub, bark brown to gray, smooth, becoming slightly furrowed and scaly, branchlets densely silky-haired, becoming sparsely pubescent. Leaves not glaucous, satiny on underside, broadly obovate, short-pointed at apex tapering to narrow base, margins entire to sparsely wavy toothed, dark green above, paler beneath with short silvery silky hairs, glossy sheen on lower surface of leaves distinguishes it from other willow species. Flowers slender, tightly flowered on short stalks appearing with leaves, scales small brown, densely hairy. Seed capsules short silvery-hairy.

Southeastern Alaska west to Kodiak Island and Cook Inlet. Along streams and shores, gravelly areas and alpine where it associates with Sitka Alder. Common. Flowers April-May. Leaves and bark of all willows contain source of natural precursor to aspirinsalicylic acid.

Salix arbusculoides (Littletree Willow) Shrub or tree 1-6 m tall, usually with several stems from same root; slender, glossy branches and sparsely velvet-haired branchlets, leaves glossy, stipules with white secretion often dried at base; pistils with white and rusty-colored hairs. Stream banks, forests openings, bogs and willow thickets at edge of alpine and arctic tundra. Common throughout interior Alaska including Lynn Canal.

Salix fuscescens (Alaska Bog Willow) Low trailing shrub with branches spreading from central woody base, rooting at nodes, branchlets reddish or yellowish brown, leaves obovate to elliptic, margins entire or partly glandular with minute teeth, leaves glossy and bright green above, pale green beneath, pistils pubescent with short rust-colored hairs. Trails in tundra moss, meadows and bogs throughout Alaska except for coastal SE Alaska and Aleutians. Found on Chichagof Island and in Glacier Bay.

Salix glauca (Grayleaf Willow) Erect shrub to 5 m tall, twigs reddish brown to grayish, leaves blades entire with a bluish-white waxy coating below. When young, they are hairy on both sides but can become nearly hairless with age, elliptic to oblong, entire or rarely minutely glandular-to minutely toothed on lower part; catkins appear with leaves. Widely distributed in open gravel areas from Juneau north and west to Alaska Peninsula and into interior Alaska.

Salix hookeriana (Bigleaf Willow) Shrub to 5 m tall with stout gray hairy twigs, branchlets with dense white woolly hairs, leaves hairy, oval to egg-shaped, margins distantly and irregularly glandular with blunt teeth, yellow green, glossy above. Sand dunes, meadows and willow thickets; Yakutat to Prince William Sound. Used for rope, fishing lines and poles for fish weirs, since they take root.

Salix lanata (Woolly Willow) Shrub to 7 m tall, branches with coarse, thick, spreading hairs; leaves elliptic, margins entire or glandular with minute teeth, upper leaves hairless or sparsely long, soft, wavy haired. Arctic and alpine thickets, on sand or gravel, spruce scrub on mountain slopes and wet meadows throughout Alaska north of Juneau.

Salix lasiandra (Western Black Willow) Shrub or small tree 1-12 m tall, twigs glossy with duck-billed shaped buds, branchlets tawny, pubescent, leaves narrowly or broadly ovate, margins glandular with blunt teeth, upper side green and glossy when mature, lower side sometimes with rust-colored hairs, becoming hairless. River banks and alluvial deposits, wet meadows in southeastern Alaska near Haines, Skagway and Yakutat; also south-central Alaska, and along Yukon River and tributaries. Also known as ***S. lucida*** spp ***lasiandra.***

Salix monticola (Mountain Willow) Shrub 1-4 m tall, branchlets yellow green, hairless, upper side of mature leaf green and hairless. Forests seeps and bogs from Juneau north to central Alaska and eastern Alaska Range.

Salix ovalifolia (Ovalleaf Willow) Prostrate creeping dwarf shrub with trailing or almost rhizomatous branches, branchlets greenish brown, hairless and glossy, leaves elliptic to round on slender petioles, upper surface green, lower surface pale green to whitish, margins entire; twigs slender, orange to dark reddish brown. Alpine tundra, moraines and sandy lake margins from southeastern Alaska to Kenai and Alaska Peninsulas and Kodiak Island. Closely related to or same as ***S. stolonifera***.

Salix planifolia (Diamondleaf Willow) Prostrate or ascending shrub 1-4 m tall, branchlets brownish, glossy, hairless to with dense whitish gray, long soft wavy hairs, leaves elliptic, margins entire to glandular with blunt teeth, upper side glossy when mature, lower with white or rust-colored hairs or hairless. Arctic and alpine tundra, subarctic and boreal forests and bogs in most of Alaska. Young leaves eaten raw; young underground shoots peeled and eaten raw. Includes ***S. pulchra***.

Salix polaris (Polar Willow) Dwarf shrub often partly subterranean, hairless, leaves obovate, entire and glossy, pistils reddish and glossy. Arctic and alpine tundra to 1700 m on rocky cliffs on Juneau Ice Field and St. Elias Range to central Alaska and Alaska Range, Bering Sea region north to Point Hope and Atka Island in Aleutians.

Salix reticulata (Netleaf Willow) Trailing shrub, branchlets greenish brown, hairless, leaves leathery, stiff, raised nerves below, rolled margins, leaf shape from oblong-obovate to nearly round, with long silky white hairs to hairless; stigma purplish, filaments pubescent. Arctic and alpine tundra, rare in boreal forests and bogs throughout Alaska. To 1500 m on nunataks of Juneau Ice Field, Admiralty Island to Glacier Bay and Lynn Canal area north throughout Alaska.

Salix rotundifolia (Least Willow) Dwarf, largely subterranean shrub to 30 cm tall, branches yellow brown, hairless, branchlets with 2-3 circular leaves, margins entire, sometimes fringed with hairs, both sides glossy, primary veins usually prominent, pistils hairless and reddish brown. Arctic and alpine tundra throughout Alaska.

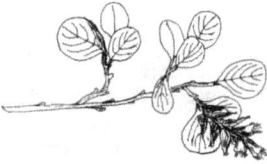

Salix setchelliana (Setchell Willow) Prostrate or semi-prostrate shrub to 25 cm tall, branches gray brown, bark loose, leathery, branchlets reddish, mature leaves leathery, narrowly obovate, entire or glandular with minute teeth, upper side lemon green, hairless, veined when mature, lower side hairless, pale yellow green; pistils brick red, hairless. Sandy beaches, glacial river gravels, glacial moraines in Alaska Range and interior river valleys. Isolated specimen found on Alsek River terrace near Yakutat. *Salix interior* and *S. prolixa* reported near Petersburg.

SANTALACEAE
SANDALWOOD FAMILY

Geocaulon lividum   **Pumpkinberry**

Rhizomatous root parasite to 30 cm tall from brown to reddish rhizome, hairless. Leaves alternate, ovate to oblong, elliptic or oblanceolate. Flowers in axils of middle to almost terminal leaves, sepals triangular, petals lacking, sepals green to reddish purple. Fruit fleshy, scarlet, solitary, one-seeded.

Bogs, alpine and arctic tundra, forests. Interior Alaska and northern SE Alaska. Flowers June-July. Fruit edible but not palatable. Leaves sometimes infected with blister rust which distorts and colors the leaves in beautiful patterns. Also known as *Comandra lividum* and Bastard Toadflax.

SAXIFRAGACEAE
SAXIFRAGE FAMILY

KEY TO GENERA OF *SAXIFRAGACEAE:*

1a. Petals lacking; sepals 4; stamens 4-8; flowers in axils of upper leaves, inconspicuous.   *Chrysosplenium*
1b. Petals usually present; sepals 5; stamens (3)5-10; flowers often showy. **(2)**
2a. Pistil with 4 carpels; stamens 5, alternating with staminodia.   *Parnassia*
2b. Pistil usually with 2 carpels, rarely with more; stamens (3)5-10; staminodia lacking. **(3)**
3a. Ovary with 2 locules, rarely with more; placentae axile. **(4)**
3b. Ovary with single locule; placentae 2, rarely 3, parietal to basal. **(5)**
    4a. Leaves leathery, broadly elliptic to obovate, margins with blunt teeth; carpels nearly distinct (at least in flower).   *Leptarrhena*
    4b. Leaves leathery or not, but if so, then not broadly elliptic to obovate; carpels united for some distance above   *Saxifraga*
5a. Petals with pinnate lateral lobes; flowers in racemes. **(6)**
5b. Petals entire; flowers in panicles or racemes. **(7)**
    6a. Calyx distinctly united, hypanthium appearing tubular, mostly over 6 mm long; plants over 30 cm tall.   *Tellima*
    6b. Calyx saucer shaped, much less than 6 mm long; plants often less than 30 cm tall.   *Mitella*
7a. Stamens 10; carpels unequal in size, one about twice as large as other; sepals white to pink.   *Tiarella*
7b. Stamens 3-5; carpels essentially equal in size; sepals green or greenish purple. **(8)**
8a. Flowers in panicles; petals white; stamens 5.   *Heuchera*
8b. Flowers in racemes; petals brownish purple; stamens 3.   *Tolmiea*

Chrysosplenium tetrandrum
Northern Water-carpet

Perennial aquatic herb 1-15 cm tall from slender almost rhizomatous stolons. Basal leaves rounded-kidney shaped to heart-shaped with a few shallow lobes. Flowers inconspicuous, nestled among leaves, usually with 4 sepals and mostly 4 stamens opposite sepals, petals lacking, sepals green or tinged purple. Fruit capsule 3-4 mm long, smooth shiny red seeds.

Juneau area north throughout Alaska. Wet areas near fresh water and streams. Occasional. Flowers June-July. Raindrops hit the shallow cups and splash out seeds; similar to"splash-cup" spore dispersal mechanism of bird's nest fungus.

Heuchera glabra   Alpine Heuchera

Perennial herb 15-60 cm tall from stout rootstock, stem base with persistent brown leaf bases and stipules. Basal leaves shiny and finely toothed, 2-lobed stipules at base of leaf petiole with a few short hairs, long-petiolated; stem leaves 1-3, reduced, with short petiole to stalkless. Flowers tiny, numerous, in loose panicle, calyx lobes broadly ovate, petals glandular with long slender claw and entire margined blade, 5 stamens, 5 triangular sepals shorter than 5 white petals. Fruit capsule with slender beak, seeds with rows of spines.

Southeastern Alaska west to Kodiak Island and Alaska Peninsula. Moist rocks, stream banks, seashore cliffs and alpine meadows along seeps. One of our most common Saxifrages. Dried roots pounded and made into a poultice to stop bleeding and promote healing. Flowers June-August.

Leptarrhena pyrolifolia Leatherleaf Saxifrage

Perennial herb 20-40 cm tall, stem base with leaves from previous years, stem sparsely glandular with one or two leaves. Leaves thick, lustrous and leathery, basal leaves ovate, sharp forward pointing teeth on margins, short stalked, hairless when young, becoming brown beneath. Flowers small, on glandular stalks rising above basal leaves, 5 ovate-oblong sepals, 5 showy linear petals, twice as long as sepals, petals white or whitish. Fruit 2 follicles bright red or purple.

Southeastern Alaska west to Aleutians. Moist alpine meadows, along creeks from sea level to alpine. Occasional to common. Flowers June-July, commonly found in fruit.

Mitella pentandra   Alpine Mitrewort

Perennial herb 20-40 cm tall from stout rhizome, rarely producing stolons. Basal leaves long stalked, heart-shaped to round and palmately lobed, stem leaves lacking. Flowers small, inconspicuous, each of 5 petals alternate to 5 stamens, and pinnatifid with 6-10 closely set, threadlike segments, sepals broadly triangular, ovary at least one half inferior, petals greenish yellow. Fruit capsules 2-3 mm long with numerous seeds.

Southeastern Alaska west to Aleutians. Along streams, moist woods, marshes, rocks and alpine meadows. Common. Flowers June-July.

Mitella nuda (Stoloniferous Mitrewort) Smaller 7-20 cm tall often stoloniferous, with 10 stamens and greenish yellow flowers, ovary about one third inferior. Moist woods, bogs and along streams in southeastern Alaska. Flowers June-August.

Mitella trifida (Three-toothed Mitrewort) 12-40 cm tall, petals white or purplish, stamens 5, ovary about one half inferior. Moist woods in coastal southeastern Alaska. Flowers June-July.

Parnassia fimbriata
Fringed Grass-of-Parnassus

Perennial hairless herb 15-30 cm tall with clasping bract near middle of flowering scape. Leaves kidney shaped, long-petiolated. Petal edges fringed along lower margins and gradually tapering into claw, fringed gland at each of 5 petal bases with stubby lobes, staminodia mostly 5-9 segments, 5 elliptic sepals with glandular teeth in margin, petals white. Fruit ovoid capsule 9-12 mm long.

Southeastern and south-central Alaska. Bogs, wet meadows, banks of streams, especially in alpine and recently deglaciated areas. Common. Flowers July-early September.

Parnassia palustris
Northern Grass-of-Parnassus

Perennial herb 10-45 cm tall, rhizome short, stout, single stem, stalkless bract below middle of scape. Leaves mainly basal, ovate, heart-shaped or elliptic, smaller than leaves of other two species. Showy petals with 7-9 veins, staminodia dilated with several slender hairs nearly as long as stamens, petals white. Fruit capsules 8-12 mm long.

Throughout Alaska. Wet meadows. Occasional. Flowers July-August. Not at all a grass.

Parnassia kotzebuei (Kotzebue Grass-of-Parnassus) Tends to be smaller than *P. palustris*, 3-20 cm tall, bract lacking or near base of scape, leaves short-petiolate or almost stalkless, much smaller and more basal, flowers inconspicuous, petals entire margined. Arctic and alpine tundra from Juneau north throughout Alaska. Flowers July-August.

KEY TO SPECIES OF *SAXIFRAGA:*
1a. Plants not bearing scape; stem leaves present below inflorescence, or inflorescence reduced to solitary flower. **(2)**
1b. Plants bearing scape; stem leaves lacking below branches of inflorescence
(branches sometimes subtended by reduced, leaf-like bracts); inflorescence with
few to many flowers. **(9)**
  2a. Leaves toothed or lobed (at least some). **(3)**
  2b. Leaves all entire (coarsely ciliate in some). **(6)**
  3a. Basal leaves leathery, tridentate, teeth minutely spiny-tipped. *S. tricuspidata*
  3b. Basal leaves not leathery, with 3-7 teeth or lobes, these not minutely spiny-
tipped. **(4)**
  4a. Bulbets present in axils of upper leaves; inflorescence with only uppermost
flower (rarely 2) developing. *S. cernua*
  4b. Bulbets lacking in axils of uppermost leaves; inflorescence 1-several flowers.
(5)
  5a. Leaves merely 3-toothed; staminal filaments almost equal to sepals; plants of
southern and southeastern Alaska. *S. adscendens*
  5b. Leaves with 3-7 distinct lobes; staminal filaments longer or shorter than sepals;
plants of various distribution. *S. caespitosa*
6a. Leaves opposite; petals pink purple to blue purple, or rarely white.
  S. oppositifolia
6b. Leaves alternate; petals white, yellow, reddish purple, or tinged with pink. **(7)**
7a. Leaf margins with bristles. **(8)**
7b. Leaves hairless, margins without bristles. *S. tolmiei*
  8a. Flowers solitary, borne stalkless on basal rosette, or on stem to 1 cm long; petals
minute, to 1 mm long. *S. eschscholtzii*
  8b. Flowers 1-several, borne on stems usually over 3 cm long; petals 3-10 mm long.
  S. bronchialis
  9a. Leaves entire. *S. tolmiei*
  9b. Leaves lobed or toothed or both (at least some). **(10)**
  10a. Leaf blades distinctly heart-shaped at base, or broader than long, or both. (11)
  10b. Leaf blades never heart-shaped, always wedge-shaped or acute at base,
longer than broad (except in *S. lyallii*). **(12)**
    11a. Leaf blades broadly lobed, each lobe with 3-more teeth.
    S. mertensiana
    11b. Leaf blades toothed or lobed but not both. *S. nelsoniana*
12a. Flowers numerous, borne in dense clusters along an elongate, spike-like
panicle; petals reddish purple at flowering; leaf blades elliptic, mostly 2-3
times as long as broad, often over 5 cm long. *S. hieracifolia*
12b. Flowers few to several or many, in open panicles (compact in *S. nivalis*); petals white to pink at flowering; leaf blades mostly not as above. **(13)**
13a. Leaves with reddish, glandular, closely interwoven and tangled hairs on lower
surface, blades with mostly 15-30 teeth; inflorescence corymbose, open;
plants of southeastern Alaska. *S. occidentalis*
13b. Leaves pubescent or hairless on lower surface, lacking reddish glandular,
densely interwoven hairs (except in *S. nivalis*), blades often with less than 15
teeth; plants broadly distributed. **(14)**
14a. Staminal filaments club shaped; capsules and ovaries various, but if reddish
purple, then branches of inflorescence with stipitate glands, or flowers more
than 10. *S. lyallii*
14b. Staminal filaments awl-shaped to linear, but if club-shaped and capsules reddish purple, then branches of inflorescence not stipitate-glandular and flowers
less than 10. *S. ferruginea*

Saxifraga bronchialis   Spotted Saxifrage

Matted perennial herb 5-15 cm tall, woody base with densely and partly overlapping leaves, flower stem glandular-pubescent with a few alternate leaves. Leaves entire margined, leathery, elliptic to oblong to flattened spoon-shaped, margins fringed with hairs. Flowers showy, sepals greenish, ovate, hairless, stamens usually longer than petals, petals yellowish, whitish or yellow-to-red spotted. Fruit two capsules with purple-black tips.

Much of Alaska except central interior valleys. Cliffs and talus slopes, from alpine tundra to open woodlands. Occasional to common. Flowers July-August. Includes *S. funstonii.*

Saxifraga ferruginea   Alaska Saxifrage

Perennial herb 10-35 cm tall with stems mostly single from thick rootstock. Basal leaves strap-like with occasional jagged teeth, tapering to broad petiole, hairless to coarse hairs, fringed with hairs on margin. Flowers showy with two kinds of petals: 3 broad ones each with two yellow-orange spots, 2 narrow petals without markings, petals white. Fruit 3-6 mm long capsule.

Southeastern Alaska west to Alaska Peninsula. Wet cliffs and wet meadows from sea level to alpine. Common. Flowers June-August.

Saxifraga lyallii   Red-stem Saxifrage

Herb 7-30 cm tall from basal rosette, flower stems leafless. Leaf blades wedge-shaped-oblanceolate to fan-shaped with 7-17 teeth or lobes, margins somewhat fringed with hairs. Flowers showy in open panicles, sepals purplish to green, lance-shaped, reflexed, filaments club-shaped, sometimes petal-like, white to yellowish. Fruit capsules 6-11 mm long.

Southeastern Alaska west to Aleutians, less common in north-central Alaska. Wet places along streams and seeps in tundra and woods. Occasional. Flowers June-July. Hybrids with *S. nelsoniana* may occur.

Saxifraga nelsoniana   Heart-leaved Saxifrage

Perennial herb with hairy, leafless, central flowering stalk 3-60 cm tall. Leaf blades round to kidney-shaped, heart-shaped at base, 9-19 teeth or lobes, each with tiny point. Showy inflorescence, calyx lobes oblong, often purplish, hairy, reflexed with age, petals twice as long as calyx lobes, stamen club-shaped, petals white to pink or rose. Fruit capsules 7-12 mm long.

Most of Alaska except central valleys. Along streams, alpine meadows and open areas. Common. Flowers May-July. Highly variable species. Leaves eaten preserved in seal oil. Also known as *S. punctata.*

Saxifraga oppositifolia   Purple Mt. Saxifrage

Matted cushion forming perennial herb with much branched stem, densely covered with old leaves and opposite leaves in 4 rows. Leaves oblong to broadly obovate, keeled, with few coarse hairs in margin. Flowers stalkless, sepals opposite and smaller than reddish violet petals. Fruit follicles with long style.

Most of Alaska except central valleys. Gravelly areas, rock outcrops, alpine. Occasional. Flowers late April-June.

Saxifraga tricuspidata   Three-tooth Saxifrage

Perennial herb 3-25 cm tall, branches of woody base clothed with overlapping leaves, flowering stems with few to several alternate leaves. Leaves leathery with three teeth at apex, linear to oblong to somewhat wedge-shaped, stem leaves similar to basal leaves, except usually entire. Flowers showy, few to several in short and broad flat topped cyme, sepals green, often tinged reddish, ovate to lance-shaped, glandular to margins fringed with hairs, spreading, stamens mostly longer than sepals, filaments awl-shaped, petals white or pale yellow. Fruit capsules 4-8 mm long.

Juneau north throughout Alaska. Cliffs, gravelly slopes and flats in tundra or woods. Flowers late May-July.

Saxifraga adscendens (Wedge-leaf Saxifrage) Perennial herb 3-10 cm tall, glandular-pubescent, basal leaves obovate with wedge-shaped base, in basal rosette, 3-5-toothed to shallowly lobed, basal leaves persistent, reddish or brownish; several stem leaves; inconspicuous flowers, calyx lobes ovate-triangular, sepals often reddish purple, petals white, about twice as long as calyx lobes. Rock crevices and glacial moraines from southeastern to south-central, central and western Alaska.

Saxifraga caespitosa (Tufted Saxifrage) Perennial 1-16 cm tall, woody base with densely overlapping leaves, flower stalk densely glandular-pubescent with 1-to several alternate leaves, basal leaves with 3-5 apical triangular to lance-shaped or narrowly oblong lobes, wedge-shaped at base, stem leaves mostly entire margined; flowers showy, white to yellowish, sepals often purplish, ovate, glandular-pubescent and margins fringed with hairs, petals 3-veined, stamens shorter than petals, longer than sepals. Tundra, gravelly ridges, cliffs and talus slopes often on limestone in much of Alaska except interior central valleys. Flowers June-August.

Saxifraga cernua (Nodding Saxifrage) Perennial 8-25 cm tall, stems with several alternate leaves, basal leaves long-petiolate, with bulbets in axils, blades 5-7-lobed, kidney-shaped, stem leaves becoming smaller, fewer lobed and shorter petiolated upwards, at least upper ones with bulbets in axils; flowers showy, solitary or rarely 2, at apex of inflorescence, others replaced by bulbets, sepals green to dark reddish purple, ovate to lance-shaped, glandular and margins fringed with hairs, petals white, 3-5-veined, stamens longer than sepals, much shorter than petals. Arctic and alpine tundra mostly in wet areas or on cliffs and talus slopes from Juneau north to northern, western and southwestern Alaska. Flowers July-August.

Saxifraga eschscholtzii (Cushion Saxifrage) Perennial 1-3 cm tall, mat-forming, flowering stems lacking or to 1 cm long, leaves oblong, alternate, densely overlapping, margins entire, fringed with hairs; flowers inconspicuous, dioecious, sepals greenish to yellowish or reddish purple, spreading to reflexed, early deciduous, pink to white, 1-veined, stamens shorter than sepals in pistillate flowers, longer than sepals in staminate flowers. Rocky slopes and cliffs in Haines area, Seward and Kenai Peninsulas north and east along Brooks Range to northeastern Alaska. Flowers in July.

180

Saxifraga mertensiana (Wood Saxifrage) Perennial 15-35 cm tall, flowering stems leafless with long soft wavy hairs, leaves with 13-17 lobes at least main lobes with 3-5 teeth, blades oval to round, margins sparsely fringed with hairs; flowers showy, lateral ones often replaced by bulbets, sepals green or purplish, lance-shaped, reflexed, white petals, stamens longer than sepals; filaments club-shaped. Wet sites, woods to alpine on coasts and islands from SE Alaska north and west to Kodiak Island. Flowers May-July.

Saxifraga nivalis (Snow Saxifrage) Perennial 5-22 cm tall, from basal rosette, flower stems leafless, glandular to with long soft wavy hairs, leaves 21-toothed, blades oblanceolate to flattened spoon-shaped, wedge-shaped at base; flowers showy in 1-3 dense spike-like clusters, sepals green to purplish, ovate to triangular, erect, petals white to pink or rarely reddish purple, filaments awl-shaped. Tundra and woods in most of Alaska except central interior valleys, Aleutians and southern Panhandle. Highly variable species. Flowers June-July.

Saxifraga occidentalis (Rusty Saxifrage) Perennial 10-25 cm tall, arising from basal rosette with short, stout rhizome; flowering stems glandular-villous, leafless; leaves mostly with 15-30 teeth, blades ovate to elliptic, wedge-shaped at base, at least younger reddish glandular-tomentose on lower surface, ciliate; flowers showy, several to many in an open flat-topped panicle, sepals green to purplish, petals white to pink, stamens longer than sepals, club-shaped, capsules 3-6 mm long. Rarely, in woods in southeastern Alaska. Flowers May-August.

Saxifraga tolmiei (Alpine Saxifrage) Perennial 2-8 cm tall, flowering stems with stalked glands, not densely woolly, basal leaves entire, oblanceolate to flattened spoon-shaped, hairless, 1-2 cilia near base, stem leaves 1-few, alternate, similar to basal leaves; flowers 1-3, (rarely more) inconspicuous, sepals green to purplish, lance-shaped, spreading, hairless, petals white, stamens mostly longer than calyx lobes, filaments club-shaped, almost petal-like. Rock crevices, talus slopes, open meadows in alpine tundra in Southeast Alaska. Flowers July-August.

Tellima grandiflora   Fringecup

Perennial herb 40-80 cm tall from stout rootstock, stems hairy. Basal leaves maple-like, hairy, stem leaves reduced. Flowers cuplike with 5 strap-like petals tips fringed, flowers scattered on spike, stamens 10, short filaments, petals white, aging red-brown. Fruit: ovate capsule, styles divergent, seeds with small rounded pimple-like structures.

Southeastern Alaska west to Alaska Peninsula and Kodiak Island. Meadows, forest edges , wet rocky areas. Occasional. Flowers July-early August. Eaten by elves to improve night vision.

Tiarella trifoliata   Foamflower

Perennial, rhizomatous herb 15-50 cm tall, slender, glandular-hairy. Basal leaves long-stemmed, divided into three segments, leaflets stalked, flowering stem with 1-2 leaves divided into 3 segments. Flowers in cyme-like panicles, stamens long, conspicuously whisker-like and well beyond 5 slender, entire, coiled petals, 5 sepals white to pink, petals white. Fruit capsule resembles sugar scoop.

Southeastern Alaska west to eastern Kenai Peninsula. Wet areas, forests, meadow, seeps, floodplains and stream sides. Common. Flowers June-September.

Includes *T. unifoliata* (Oneleaf Foamflower) which has simple rather than compound leaves that are maple-like and shallowly 3-5-lobed.

Tolmiea menziesii   Youth-on-age

Slender perennial herb 40-70 cm tall, glandular or with rough hairs or bristles, from stout rootstock, stem hairy. Leaves maple-like, narrowly 5-7-lobed, lobes irregularly toothed with coarse, scattered white hairs. Flowers borne in racemes, cylindrical flower tube tipped by 4 petals surrounded by calyx with 3 equal teeth and 2 short lateral ones, petals brownish purple. Fruit capsules 9-14 mm long, valves equal.

Southeastern Alaska. Moist deciduous woods and along streams. Occasional. Flowers June-August. Small buds at the base of leaf blades develop "daughter plants" giving the plant its common name.

SCROPHULARIACEAE
FIGWORT FAMILY

KEY TO GENERA OF *SCROPHULARIACEAE*:
1a. Anther bearing stamens normally 2.   *Veronica*
1b. Anther bearing stamens normally 4. **(2)**

2a. Staminal filaments 5 or 5th represented by gland at base of corolla, 4 anther-bearing, fifth sterile and lacking an anther; leaves opposite. **(3)**
2b. Staminal filaments 4, all anther-bearing; leaves alternate, basal, whorled, or opposite. **(4)**

   3a. Plants annual; corolla 7 mm long or less, middle lobe of lower lip deeply concave and enclosing stamens; sterile stamen reduced, minute and gland-like.   *Collinsia*
   3b. Plants perennial; corolla over 7 mm long, middle lobe of lower lip not enclosing stamens; sterile stamen elongate, bearded.
                                Penstemon
4a. Leaves opposite. **(5)**
4b. Leaves alternate or basal (whorled in some *Pedicularis* species). **(7)**
   5a. Corolla lobes of upper lip spreading, not developed into a hood or a beak.   *Mimulus*
   5b. Corolla lobes of upper lip inconspicuous, developed into a hood which arches over stamens. **(6)**
   6a. Leaves lance-shaped, with sharp forward pointing teeth and rough hairs or bristles; calyx somewhat inflated in flower, much so in fruit.   *Rhinanthus*
   6b. Leaves ovate to orbicular, coarsely round to sharp teeth to lobed, not roughened-hairy; calyx not much inflated.
                                Euphrasia
7a. Leaves all basal; corolla small and inconspicuous, essentially regular; plants diminutive aquatics, growing in mud.   *Limosella*
7b. Leaves along stem, at least some (except in some *Pedicularis*); corolla conspicuous, irregular; plants not aquatic but some growing in mud. **(8)**
8a. Leaves toothed or dissected, or rarely pinnately compound, mostly basal, stem leaves often poorly developed or lacking; pollen sacs similar.   *Pedicularis*
8b. Leaves entire, or pinnately cleft, stem leaves well developed, basal ones reduced in size or lacking, pollen sacs unequal.
                                Castilleja

KEY TO SPECIES OF CASTILLEJA:
1a. Lower lip of corolla less than one-fifth as long as the galea. **(2)**
1b. Lower lip of corolla at least one-forth as long as the galea. **(3)**
  2a. Bracts of the inflorescence yellow to yellow-orange; calyx lobes and bracts obtuse to rounded at the tip.
        C. unalaschcensis
  2b. Bracts of the inflorescence red to crimson or red orange (rarely yellowish); calyx lobes and bracts obtuse to sharp at the tip.   *C. miniata*
  3a. Leaves with 1-3 pairs of prominnent lobes; corolla mostly 12-17 mm long.   *C. parviflora*
  3b. Leaves entire; corolla mostly less than 17 mm long. *C. pallida*

Castilleja miniata   Red Paintbrush

Perennial herb 20-80 cm tall, stem simple or somewhat branched above, hairy especially above. Leaves lance-shaped, acute, entire margin or upper leaves occasionally lobed. Inflorescence hairy, bracts more or less lobed, acute, scarlet tipped, calyx lobes lance-linear, upper corolla hairy, as long as tube with red margins, lower lip green, teeth united or nearly so, petals red. Fruit capsule 9-10 mm long.

Southeastern Alaska west to northern Alaska Peninsula (except Kodiak Island) and interior Copper River drainage. Meadows, salt marshes and beach meadows. Common. Flowers June-August. Red Paintbrushes are highly variable; a complex species including *C. hyetophila* and *C. chrymactis*.

Castilleja pallida   Green Paintbrush

Hairless to densely hairy perennial herb 15- 50 cm tall. Leaves simple or uncommonly upper ones toothed, bracts and lobes greenish, yellowish, or cream. Galea has distinct teeth, flower color green. Fruit capsule 7-10 mm long.

Most of Alaska from Lynn Canal and Glacier Bay north to interior Alaska. Streams and meadows. Occasional. Flowers June-August. Also known as *C. caudata.*

Castilleja parviflora   Alpine Paintbrush

Perennial herb 10-30 cm tall from woody base, turning black when dried, hairless or with sparse, long, soft wavy hairs. Leaves lance-shaped except lower leaves which have 1-2 pairs of lateral lobes, bracts lance-shaped to ovate, 3-5 lobed, sometimes entire, rose pink to crimson. Lateral lobes distinct for 3-6 mm, also rose pink to crimson, galea with distinct teeth, petals green. Fruit 2-celled capsules 8-11 mm long.

Southeastern to south-central Alaska. Subalpine to alpine meadows. Occasional. Flowers July-August. Paintbrushes are often partially parasitic on roots of other plants.

Castilleja unalaschensis   Yellow Paintbrush

Perennial herb 20-80 cm tall, stems hairless to sparsely hairy below, hairs more numerous on flower. Leaves lance-linear to lance-shaped, sharp-tipped, bracts lance-shaped to ovate or wedge-shaped, entire or with 2 broad, lateral greenish to bright yellow lobes. Lateral lobes of calyx distinct for 5-10 mm, greenish to bright yellow, teeth connate or nearly so, petals green. Fruit capsule 8-12 mm long.

Southeastern Alaska west to Aleutians. Subalpine meadows, open woods and tidal flats. Common. Flowers June-early August.

Collinsia parviflora   Blue-eyed Mary

Annual herb 5-40 cm tall, often found growing in masses, stem simple or branched from thin root. Lower leaves small, flattened spoon-shaped to round, stalked, upper narrowly elliptic to linear, stalkless. Flowers long-stalked, lower solitary in axils, upper clustered, corolla tube abruptly bent at wide angle, petals blue with whitish upper lip. Fruit ellipsoid capsule.

Mainland of southeastern Alaska except Yakutat. Moist open places, ditches. Occasional, common near Haines. Flowers June.

Euphrasia arctica　　　　Arctic Eyebright

Parasite on roots of various plants, to 30 cm tall, stem usually simple and hairy. Leaves opposite, mostly 1-4 pairs above cotyledons, prominently toothed. Flowers inconspicuous, on leafy, bracted terminal spikes, calyx teeth triangular to lance-shaped, petals whitish. Fruit capsule 3.5-5 mm long.

Juneau north throughout Alaska south of 65th parallel. Open areas, wet meadows and roadsides. Occasional. Flowers late July-August. Includes *E. mollis* and *E. disjuncta*.

Limosella aquatica　　　　Mudwort

Small aquatic herb 2-8 cm tall, stemless, hairless perennial with fibrous roots forming cushion-like mats. Leaves narrow, simple, entire margined, long-stalked; tiny flowers borne singly on elongate leafless peduncle, calyx with 5 spreading lobes. Fruit 2-loculed capsule 2-3.5 mm long.

Southern half of Alaska in widely disjunct locations. Mud banks and shallow water. Uncommon, probably much overlooked. Mudwort leaves usually have a flattened, spoon-shaped blade. Flowers July-September.

Mimulus guttatus　　　　Yellow Monkey-flower

Annual or perennial herb 10-70 cm tall. Lower leaves ovate to oblong entire or with sharp, forward pointing teeth, stalked; upper leaves ovate to round, stalkless. Racemes with pedicels shorter than corolla, corolla lower lip much longer than upper, petals yellow often marked with red or purple in throat. Fruit long, obtuse capsule.

Southeastern Alaska west to Aleutians, also in much of Alaska south of the 65th parallel. Wet areas, along streams and on rocky slopes. Common. Flowers late June-August. Monkey-flowers have 2-lobed stigmas that close when touched.

Mimulus lewisii (Purple Monkey-flower) Perennial 30-100 cm tall, sticky and softly hairy, leaves paired and clasping stem, flowers pink-purple marked with yellow and not spurred; capsules oblong, 10-20 cm long with numerous seeds. Moist sites, streams and woods in extreme southeastern Alaska; reported near Hyder and on Baranof and Douglas islands where apparently rare. Flowers July-August. Flowers July-August.

186

KEY TO SPECIES OF *PEDICULARIS*:

1a. Flowers cream to yellow, sometimes tinged with pink, red orange, or red purple. **(2)**

1b. Flowers rose pink to pink purple or purplish red, often variously marked with other colors. **(4)**

    2a. Plants annual or biennial (short-lived perennial), stems branching; bracts within inflorescence almost equal to flowers or much longer.   *P. labradorica*

    2b. Plants perennial, stems simple or rarely branching from base; bracts within inflorescence mostly much shorter than flowers. **(3)**

    3a. Flowers cream-colored (often tinged with pale pink or red purple), 25-40 mm long; racemes usually in heads, few-flowered.   *P. capitata*

    3b. Flowers pale to bright yellow, 12-20(23) mm long; racemes usually elongate, mostly several- to many-flowered.   *P. oederi*

4a. Stems with one or more sets of whorled leaves.   *P. verticillata*

4b. Stems with leaves alternate (some rarely almost opposite), or leaves all basal (and still alternate). **(5)**

5a. Galea prolonged apically into prominent beak, beak 2 mm long or more; plants of southeastern Alaska.   *P. ornithorhyncha*

5b. Galea not prolonged into beak, blunt apically or with a pair of almost terminal teeth; plants widely distributed. **(6)**

6a. Plants annual or biennial (short lived perennial?); stems often branched above base, or if simple then inflorescence often much elongate with several apparent internodes below few-flowered, almost capitate apex.   *P. parviflora*

6b. Plants perennial; stems simple or branched from woody base; inflorescence seldom much elongated.   *P. sudetica*

Pedicularis oederi   **Oeder Lousewort**

Perennial herb 10-70 cm tall with thick stem, hairless at base from short woody base with several thick, yellowish-white spindle-shaped roots. Leaves thick, dark green, deeply sharp-toothed, fernlike. Tip of upper lip of corolla brownish-red with 2 red spots, lower lip 3 lobed, petals yellow. Capsule lance-shaped to oblong, style protruding.

Most of Alaska. Alpine and arctic tundra, high elevation. Common. Flowers July-August.

Pedicularis parviflora   Small-flowered Lousewort

Annual or short-lived perennial herb 20-70 cm tall. Leaves alternate, mostly on stem, stalkless or almost stalkless, pinnately lobed to pinnatifid, lobed margins with round or blunt teeth, bracts similar to leaves but reduced upwards. Inflorescence elongate with several apparent internodes, calyx expands and partially encloses capsule in fruit, cleft above and below, 2 lateral lobes irregularly toothed, galea straight to slightly curved, not beaked or with teeth almost at apex, petals purple or bicolored. Fruit capsules 8-17 mm long.

Southeast Alaska north to south-central and west-central Alaska. Wet meadows, bogs, low to mid elevation. Common. Flowers July. Includes *P. macrodonta*.

Pedicularis verticillata   Whorled Lousewort

Perennial herb 8-40 cm tall, inflorescence densely woolly, usually 1-2 whorls of leaves below bracts. Leaves short to long stalked, deeply pinnatifid, lobes toothed to incised. Inflorescence whorled in heads to elongate, upper lip (galea) inflated, somewhat transparent, 5-toothed with entire teeth, galea slightly arched, not toothed or beaked, latter containing two seeds, petals rose-pink. Capsules 10-15 mm long.

Most of Alaska. Moist to dry tundra, woods and beach meadows, mostly high elevation. Occasional. Flowers July-August.

Pedicularis capitata (Headed Lousewort) 5-16 cm tall, stems with simple sparse long soft wavy hairs, leafless or 1-2 leaves, leaves alternate, long-petiolated, pinnately once compound, leaflets lobed or dissected, bracts similar to leaves or less dissected; inflorescence in heads, 1-8-flowered, calyx 5-toothed, margins with round or blunt teeth apically, corolla cream, often tinged pink or red purple, galea strongly arched with pair of teeth almost at apex but not beaked. Moist or dry tundra in most of Alaska.

Pedicularis labradorica (Labrador Lousewort) Perennial 5-40 cm tall, sparse to dense long, soft wavy hairs commonly in lines below leaf bases, leaves alternate, winged petioles, lobed to pinnatifid, lobes toothed to incised, bracts similar to leaves but reduced, less deeply lobed upwards; inflorescence in heads to elongate, often with several internodes, calyx 2-3-lobed, lateral teeth appearing as single tooth near upper edge of tube, corolla yellow, often marked with orange or red orange, galea slightly arched, somewhat beaked and toothed just below apex. Alpine tundra and woodlands from Juneau north throughout most of Alaska except Aleutians.

Pedicularis ornithorhyncha (Birds-beak Lousewort)
Perennial 5-30 cm tall from fibrous rooted stem base,
leaves alternate or mostly basal, long-stalked, pinnatifid
to pinnately compound, lobes deeply cleft and often again
toothed, bracts similar to leaves but often much reduced;
inflorescence mostly in heads, lower internodes often
apparent, calyx 5-toothed with sharp, forward pointing
apical teeth, corolla purplish, galea strongly arched and
prolonged into conical beak. Alpine meadows of southeastern Alaska.
Flowers July-August. Many of the Lousewort stems and roots are edible
but should not be collected since they are often only locally common.

Pedicularis sudetica (Sudet Lousewort) Perennial
4-50 cm tall, stem leaves alternate or all basal, long-stalked,
pinnately lobed or pinnatifid, lobes toothed and ultimate
segments again toothed, bracts variable, usually resembling
leaves, reduced upwards; inflorescence in heads to elongate,
flowers spirally arranged, calyx 5-toothed, teeth entire to
margins with round or blunt apical teeth, corolla rose-pink,
pink-purple, or bicolored, galea moderately arched, dark
purple, lower lip pink, often spotted, beaked, toothed just below apex.
Moist to dry tundra and woods in most of Alaska. Flowers June-July.
Variable species. Young shoots boiled in soup.

Penstemon serrulatus   Beardtongue

Perennial, woody below, 20-70 cm tall,
conspicuously hairy inflorescence, all leaves on stem,
opposite with sharp forward pointing teeth, stalkless or
nearly so, flower tubular, corolla blue to purple. Fruit
capsules 5-8 mm long.

Stream banks and other moist sites. Extreme
southeastern Alaska (Hyder). Flowers July-August.
Penstemon is Latin meaning 5 stamens; this genus
has four fertile and one sterile stamen. Also known as
Coast Penstemon.

Rhinanthus minor   Rattlebox

Minutely hairy annual herb 10- 80 cm tall with
leafy stems. Leaves opposite, simple, saw-toothed,
lance-shaped. Flowers showy, sepals fused, galea with
pair of broad almost apical teeth, petals yellow. Fruit
flattened spherical capsules; numerous winged seeds.

Southeastern Alaska west to Aleutians north to
62nd parallel. Meadows, beaches and disturbed areas.
Common. Flowers July-August. Seeds make a rattling
sound inside the capsule when mature. Also known as
R. crista-galli.

Veronica americana   American Brooklime

Perennial herb 10-70 cm long, stems reclining at base, rooting at nodes, hairless. Leaves opposite, short stalked, obscurely to sharply toothed, lance-shaped to narrowly oblong. Racemes emerge from axils, 10-25-flowered, petals blue. Capsules round, slightly notched, seeds numerous.

Southeastern Alaska west to Aleutians north to 65th parallel. Along streams and in open wet areas. Occasional. Flowers July-middle August. Leaves are edible. Brooklimes used to treat urinary and kidney ailments and as a blood purifier. Also known as *V. beccabunga* ssp *americana.*

Veronica wormskjoldii   Alpine Speedwell

Perennial from shallow rhizome 7-30 cm tall from simple stem with sparse to dense spreading hairs. Leaves mostly opposite, elliptic to oval, all on stem, slightly toothed, becoming black in drying. Flowers in terminal racemes, upper flower bracts usually alternate, petals blue to violet. Fruit a hairy-glandular capsule, broadly notched.

Southeastern Alaska west to Aleutians north to 65th parallel. Moist to dry meadows in alpine and arctic tundra. Uncommon. Flowers June-August. Includes *V. stellerii.*

Veronica serpyllifolia (Thyme-leaf Speedwell) Perennial 10-30 cm tall, stems creeping or reclining at base, sparsely to densely hairy, leaves short-stalked to stalkless, entire or inconspicuously toothed, hairless or nearly so, blade lance-ovate to oval, racemes terminal on main or lateral stems, corolla blue to white, capsules heart-shaped, notched. Moist soil, in seeps, swamps, or along streams on islands and coasts of southern Alaska. Flowers June-July.

190

URTICACEAE
NETTLE FAMILY

Urtica dioica   Stinging Nettle

Perennial herb 1-3 m tall from strong, spreading rhizomes and armed with stinging hairs. Leaves opposite, lance-shaped to ovate, coarsely toothed, apically pointed. Flowers inconspicuous in axillary spike-like cymes, perianth 4-lobed, petals greenish. Fruit flattened lens-shaped achene.

Most of Alaska south of 65th parallel except Aleutians. Meadows, moist, open areas, in abundance among fern, alder, or willow on steep mountain slopes. Common. Flowers July-August. The hollow stinging hairs arise from a gland producing formic acid that causes an intense burning or numbing sensation. Young stems and leaves can be eaten if cooked and is an excellent spring tonic. Boiled young buds make excellent soup; dried leaves used for tea. Nettles contain omega 3 essential fatty acids and is an important source of fiber; stems gathered in October, dried and outer fibers spun into twine for fishing nets, blankets, etc. Tlingit people made red dye by boiling nettle stems and leaves in urine. Roots used as arrow poison. Juice from plant or alcohol helps prevent stinging sensation.

VALERIANACEAE
VALERIAN FAMILY

Valeriana sitchensis   Sitka Valerian

Perennial 30-120 cm tall from stout rhizome, 4-5 nodes below bracts of inflorescence. Mainly pinnate stem leaves, lower pair reduced, middle stalked, pinnatifid with large, terminal segment, upper stalkless, reduced. Flowers tubular with 1-sided bulge, inflorescence compact, spreading in fruit, lobes much shorter than somewhat swollen tube, stamens protruding, petals white or pinkish tinged. Achenes flattened, 1-nerved on one side, 3-nerved on other.

Southeastern Alaska to Kenai Peninsula. Moist meadows, alpine. Common. Flowers late June-August. Tlingit people applied crushed roots to mothers nipples to wean children and rubbed roots on sore muscles.

VIOLACEAE
VIOLET FAMILY

KEY TO SPECIES OF *VIOLA*:

1a. Petals white, at least on inner surface though often purple veined.　　*V. renifolia*

1b. Petals purplish, violet, lavender, pink, or yellow. **(2)**

2a. Petals yellow. **(3)**

2b. Petals purplish, violet, lavender, or pink, not yellow. **(5)**

　　3a. Aerial stems stoloniferous, rooting at nodes; flowers from near base of leafy stem; plants from extreme southeastern Alaska.　　*V. sempervirens*

　　3b. Aerial stems not stoloniferous; flowers from near tip of an erect leafy stem; plants of broad distribution. **(4)**

　　4a. Leaf blades 1-2.5(3) cm broad, kidney-shaped, apex rounded or shallowly notched (rarely acute).　　*V. biflora*

　　4b. Leaf blades 3-6 cm broad, orbicular to heart-shaped, apex acuminate to acute or with broad blunt cusp.　　*V. glabella*

5a. Plants with slender elongate stolons, leaves arising from stolons.　　*V. palustris*

5b. Plants without stolons, leaves arising from erect stems or from apex of thickened rhizome. **(6)**

6a. Head of style bearded; plants usually with elongate aerial stems, usually hairy; spur slender, to half as long as blade of lowest petal.　　*V. adunca*

6b. Head of style hairless (except in *V. langsdorffii*); plants stemless or short stemmed, hairless or minutely pubescent on upper surface; spur thick, much less than half as long as blade of lowest petal. **(7)**

7a. Leaves pubescent on upper surface; petals hairless.　　*V. selkirkii*

7b. Leaves hairless or nearly so on upper surface, lateral petals bearded.　　*V. langsdorffii*

Viola glabella　　　　　　　　Stream Violet

Herb 5-30 cm tall from thick, horizontal rhizome, flower bracts near middle or on upper scape. Aerial stem leaves much reduced, opposite, short stalked; basal leaves with long stalks, kidney-or heart-shaped. Flowers small, petals yellow on both sides, lower ones united, with purple veins, lateral pair bearded, style bearded, petals yellow. Capsule explosively splitting.

Southeastern Alaska west to Kodiak Island and Alaska Peninsula. Wet areas in forests and along streams. Common. Flowers late April-July. Tea good for colds and sore throat; leaves steamed and eaten.

Viola langsdorffii   Alaska Violet

Small perennial to 15 cm tall from thick rhizome, aerial stems poorly developed, flower scape bracts near middle of stalk. Leaves heart to kidney shaped with long leaf stalks. Flowers large, lower petals united and white at base, lateral pair with long hairs, style bearded or hairless, petals dark bluish purple. Capsule explosively splitting.

Along coast and on islands from southeastern Alaska west to Aleutians, west-central and interior south-central Alaska. Moist meadows, along streams and alpine snow beds. Common. Flowers late June-early August. Young leaves and flowers eaten raw, cooked, or dried for tea. Violets worn around the neck alleged to prevent drunkenness.

Viola palustre   Marsh Violet

Perennial from long, thin rhizomes and creeping stolons but without aerial stems. Flower scape bracts well above to well below middle. Leaves smooth and hairless on both sides, shiny especially below, kidney to heart shaped, margins toothed. Flowers on long stalks often longer than leaves, lateral pair of petals bearded to hairless, anthers orange, petals lilac to almost white, lower three petals with darker veins. Capsule, greenish, explosively splitting.

Most of Alaska. Wet meadows, stream sides, open areas and bogs, low to subalpine elevation. Common. Flowers May-July. Young leaves and flower buds eaten raw in salads and as potherb; slowly dried leaves good for tea. Also known as *V. epipsila*.

Viola adunca (Western Dog Violet) Perennial to 10 cm tall from short to long rhizomes, usually stemless in early season, aerial stems develop as season advances, leaves heart-shaped, margins with fine, round or blunt teeth, often brown-dotted, hairy, especially in margin and on veins below (sometimes hairless); petals blue to violet, lateral pair bearded, flower on leafless peduncle with 2 bracts borne above middle, style bearded. Moist sites in thickets and meadows from southeastern Alaska west to Kenai Peninsula and Kodiak Island. Flowers May-July. Flowers and leaves used as laxative and to relieve coughs and lung congestion.

Viola biflora (Two-flower Violet) Small perennial, flower on leafless peduncle with minute bracts above middle, flowers yellow to pale yellow with brownish-violet stripes, beardless with short spur and hairless style. Moist tundra, scree slopes and woods from Sitka north and west to south-central Alaska. Flowers June-July.

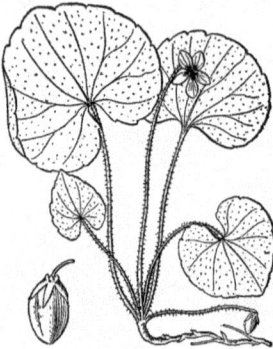

Viola renifolia (White Violet) Leaf blades heart-shaped to round to kidney shaped, margins with round or blunt teeth; flower scape bracts usually above middle of stem, petals white, lower three with purple veins, spur very short, style and petals hairless. Fertilization occurs in some flowers while still closed. Moist woods and adjacent meadows from southeastern to south-central Alaska. Flowers July-August.

Viola sempervirens (Redwoods Violet) Evergreen perennial, smaller than ***V. biflora*** and ***V. glabella***; rhizomatous, stoloniferous aerial stems on short stalks, flower scape bracts usually above middle, petals yellow, lower three with purple lines, lateral pair bearded. Otherwise very similar to ***V. biflora***. Found rarely, in woods in extreme southeastern Alaska and Prince William Sound. Flowers April-June.

Viola selkirkii (Selkirk Violet) Perennial from short to elongate somewhat thickened rhizomes, leaves hairy above especially along veins, hairless below, heart-shaped-ovate with round to sharp teeth on margin; flower on leafless peduncle with bracts mostly above middle of stem, petals violet, not bearded, with thick spur. Moist sites in woods and bogs from southeastern Alaska west to Aleutians.

MONOCOTYLEDONEAE
THE MONOCOTS

ARACEAE
ARUM LILY FAMILY

Lysichiton americanum   Skunk Cabbage

Perennial herb 30-150 cm tall, stemless from stout erect rhizome. Leaves large to 1 m long, oval, thick, hairless, in a basal cluster. Bears numerous tiny green flowers on thick, fleshy spike at end of a club-like stalk (spadix). The spadix is partially surrounded by large, bright yellow hood-like spathe which along with the lower stalk and roots, is the source of the skunk-like odor; flowers appear before leaves. Fruit berry-like and pulpy, green to reddish, embedded in fleshy flower spike.

Southeastern Alaska west to Kenai Peninsula and Kodiak Island. Open, boggy forests to edges of bogs and beaver ponds, marshes and along streams. Common. Flowers April-early June; at mid elevation, it flowers just after snow melts in June and July. Large, flat, water repellent leaves called "Indian waxed paper" and widely used to line steam-cooking pits and berry baskets. Leaf can also be used to make a temporary drinking cup and dipper by folding it in half from top to bottom and pulling edges back to lower end and holding them together with stem as a handle. Flowers attract gnats for pollination by producing fetid odor like that of decaying meat, odor lasts a short time during peak flowering. CAUTION: Contains calcium oxalate crystals which embed in mucous membranes causing irritation and a burning sensation; prolonged cooking and storage necessary to eliminate crystals before plant can be eaten. Bears dig up and eat roots in spring; deer and moose eat flowers and leaves.

In the days before salmon, Native Americans had only plants to eat, including skunk cabbage root which is very hot and peppery. The skunk cabbage decided to help the people and caused the first salmon run. As a reward, the skunk cabbage was given an elk skin blanket and a war club which he has kept to this day.

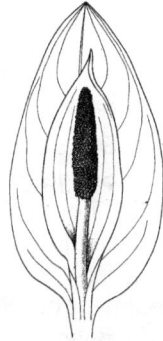

CYPERACEAE
SEDGE FAMILY

KEY TO GENERA OF *CYPERACEAE:*
1a. Flowers all unisexual; fruit naked or enclosed. *Carex*
1b. Flowers perfect, or perfect and staminate; fruit naked. **(2)**

2a. Perianth bristles many, long-silky; inflorescence often appearing like a tuft of cotton. *Eriophorum*
2b. Perianth bristles few to several; inflorescence not at all like a tuft of cotton (except in *Scirpus hudsonianus*). **(3)**

3a. Fertile flowers or achenes solitary (rarely 2) in each spikelet, lower scales of spikelet empty. *Rhynchospora*
3b. Fertile flowers or achenes few to several in each spikelet (rarely solitary), lower scales usually fertile. **(4)**

4a. Involucral leaves lacking; style base persistent as a tubercle at achene apex; spikes solitary. *Eleocharis*
4b. Involucral leaves present (1 or more); style base deciduous, tubercle lacking; spikes mostly more than 1, though solitary in some. *Scirpus*

Carex spp.   Sedges

Sedges form the largest and most complex group of plants in this field guide. Included here are the most common and distinct species. Definitive identification requires careful study of fruit morphology usually requiring the aid of a dissection microscope. For a more complete treatment of *Carex,* the reader should consult a technical reference on Sedges.

Carex anthoxanthea   Yellow-flowered Sedge

Rhizomatous, loosely tufted, 5-35 cm tall. Leaf sheaths greenish or whitish, blades of lower sheaths short or lacking, elongated upwards, 1-2 mm broad, flat to somewhat channeled, lowermost bract 0.5-1.5 cm long, blade (when present) linear-spatula-shaped. Spike solitary, narrowly cylindrical, pistillate scales almost equal to or longer than perigynia, lance-shaped to elliptic, dark brown with green to straw-colored centers, perigynia 4 mm long, 1.3-1.6 mm broad, narrowly elliptic, tapering to both ends, veined dorsally and ventrally, yellowish-green to straw-colored, hairless, stigmas usually 3.

Coast and islands of southern Alaska. Bogs and wet meadows. Common in southeastern, occasional in south-central Alaska. Flowers July-August.

Carex aquatilis   Water Sedge

Often forms large clumps or tuft from long, stout, cord-like rhizome, 15-80 cm tall. Sheaths mostly reddish to brown, blades on lower sheaths lacking, elongating above, 2-5 mm broad, flat or channeled at base, blade leaf-like. Inflorescence 5-25 cm long, erect or nodding, staminate spikes 1 or 2, slender, 1-5 cm long; pistillate or androgynous spikes 2-4, sessile or short-peduncled, 1-4 cm long; bracts leaf-like, lower exceeding the culm; scales obtuse to acuminate, blackish or reddish-brown, 1-nerved with light center, perigynia 2-3 mm long, 1.2-1.8 mm broad, appressed-ascending, obovate, hairless, short-stalked, veinless, greenish, beak 0.1-0.3 mm long, not toothed; stigmas 2, achenes lenticular.

Throughout Alaska. Bogs, meadows, stream banks and shallow ponds, low to high elevation. Common. Flowers June-August.

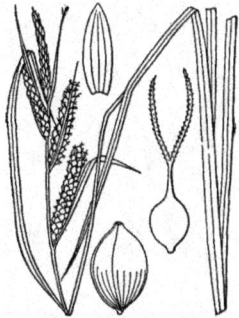

Carex gmelinii   Gmelin's Sedge

Densely tufted, 10-90 cm tall with short runners, culm coarse, curved at top. Sheaths dark brown or purplish, blades of lower sheath short or lacking, elongating upwards, 2-5 mm broad (rarely less), flat, margins rolled under or sometimes channeled below, lowermost bract 2-8 cm long, sheathless, blade leaf-like. Inflorescence 2-8 cm long, erect to nodding, spikes 3-6, terminal, stamens and pistil united in one column, lateral pistillate, upper more or less congested and commonly short-pedunculate, lower separated by 10-35 mm and with peduncles 6-15 mm long or more, ascending to erect; pistillate scales mostly longer than perigynia, elliptic to ovate, with ciliate awns 1-4 mm long or tipped with sharp, rigid point, purplish black with narrow translucent margins and lighter median; perigynia 4-5 mm long, 1.8-2.2 mm broad, lance-elliptic to ovate, appressed-ascending, wingless, hairless, short-stipitate, veined on both sides often lightly so, yellowish brown, beak 0.2-0.3 mm long, purple tipped, minutely doubly dentate; stigmas 3.

Coastal southeastern Alaska west to Aleutians and western Alaska. Tidal marshes, sandy saline shores. Common. Flowers June-July.

Carex lenticularis   Lenticulate Sedge

Densely tufted, 10-80 cm tall, rhizomes lacking or short to elongate. Leaf sheaths straw-colored to reddish-brown, blades of lower sheaths short or lacking, elongating upwards, 1-3 mm broad, flat or channeled, lowermost bract 2-20 cm long, sheathless, blades leaf-like. Inflorescence 3-14 cm long, erect or nearly so, spikes 3-6, terminal staminate or rarely gynaecandrous, lateral pistillate, all congested or lower separated by 6-30 mm or more, peduncles 3-30 mm long; pistillate scales commonly shorter than perigynia, elliptic to oblong, obtuse or somewhat acute, purplish brown to purplish black with thin translucent margins and straw-colored to greenish median; perigynia 2-3 mm long, 1.2-1.5 mm broad, ovate to elliptic, spreading-ascending, wingless, hairless, with a stipe 0.3-0.6 mm long, veined on both faces, greenish to straw-colored or brownish mottled, abruptly beaked, beak 0.1-0.3 mm long, truncate; stigmas 2.

Coastal southern and south-central Alaska. Bogs, wet meadows, gravel bars and shores from low elevation to subalpine. Common. Flowers June-August. Also known as *C. kelloggii*.

Carex limosa   Mud Sedge

Rhizomatous from yellowish, felt-like roots, 20-60 cm tall. Leaf sheaths purplish to brown, blades of lower sheaths short or lacking, developed upwards, 1-2 mm broad, channeled, lower most bract 2-7 cm long, erect to nodding. Flower spikes 2-4, terminal staminate, 15-27 mm long, lateral pistillate, separated by 25-50 mm or more, on threadlike peduncles 1.5-5 cm long, drooping; pistillate scales almost equal to perigynia, ovate to elliptic, acute to tipped with a sharp rigid point, brownish with lighter base and greenish to straw-colored median; perigynia 3-4 mm long, 1.8-2.2 mm broad, ovate to obovate, ascending, wingless, hairless, short-stipitate, veined on both sides, bluish-green to straw-colored, beak 0.1-0.2 mm long, truncated; stigmas 3.

Much of Alaska. Bogs and shallow ponds. Common. Flowers July-August.

Carex lyngbyei   Lyngby's Sedge

Rhizomatous and tufted, 20-90 cm tall with runners that take root, culms triangular, smooth, longer than leaves. Leaf sheaths purplish brown, blades of lower sheath lacking, developed upwards, 3-8 mm broad or more, flat with rolled margins, lowermost bract 8-30 cm long, sheathless, blade leaf-like. Inflorescence 7-25 cm long, erect to nodding, spikes 4-6, upper 1-3 staminate, lower pistillate (or some androgynous), separated by 15-105 mm or more, with peduncles 1.5-9 cm long, at least lowermost commonly drooping; pistillate scales longer than perigynia, lance-attenuate, attenuate to acute, purplish brown to purplish black with lighter median; perigynia 2.5-3.5 mm long, 1.5-2.5 mm broad, ovate to oval, ascending, wingless, hairless, ovate to oval, almost stipitate, obscurely veined on both sides, greenish to straw-colored, beak 0.1-0.2 mm long, truncate; stigmas 2.

Southeastern Alaska west to Aleutians and western Alaska. Tidal flats, wet meadows and bogs. Common. Flowers late May-early August. A high protein favorite food of geese, trumpeter swans, bears and other wildlife.

Carex macrochaeta   Long-awned Sedge

Rhizomatous, loosely tufted 10-90 cm tall. Sheaths purplish brown (at least below), blades of lower sheath short or lacking, elongating upwards, 2-5 mm broad, flat with rolled margins, lowermost bract 2.5-16 cm long, tubular-sheathing, sheath 3-10 mm long or more, blade leaf-like. Inflorescence 3-15 cm long, more or less nodding, spikes 2-5, terminal staminate, lateral pistillate, separated by 12-70 mm and with slender peduncles 1-8 cm long, erect or more commonly spreading to nodding; pistillate scales longer than perigynia, oblong to obovate or lance-shaped, awned (awn often over 3 mm long), purplish black to black with more or less thin, translucent margins and straw-colored to greenish median; perigynia 3-5 mm long, 1-2 mm broad, lance-shaped to elliptic, appressed-ascending, wingless, hairless, almost stipitate, finely veined, straw-colored or greenish, sometimes purplish spotted, beak 0.1-0.3 mm long, apex truncate or nearly so; stigmas 3.

Southeastern Alaska west to Aleutians and coastal western Alaska. Meadows, sandy beaches, bogs and alpine. Common. Flowers June-August. Important mountain goat forage.

199

Carex mertensii — Mertens' Sedge

Densely tufted from stout vertical rhizomes, 30-120 cm tall. Leaf sheaths purplish to brownish purple, blades of lower sheaths short or lacking, elongating upwards, 3-10 mm broad, flat with rolled margins, lowermost bract 8-27 cm long, sheathless or short-sheathing, blade leaf-like. Inflorescence large and showy, 5.5-18 cm long, more or less nodding, spikes 4-8, all gynaecandrous, terminal conspicuously so, lower separated by 15-50 mm or more, and with slender peduncles 2-5 cm long; pistillate scales equaling or shorter than perigynia, lance-elliptic, acute to shortly awned, purplish brown with thin translucent margins and lighter median; perigynia 3.4-5 mm long, 2.5-3.5 mm broad, ovate to oval, appressed-ascending, flattened, membranous, wingless, hairless, sessile, 2-veined, whitish with greenish to straw-colored apex, beak 0.2-0.5 mm long, with a shallow notch at apex, whitish to purplish tipped; stigmas 3.

Coastal southern and southeastern Alaska (rare in Interior). Open areas and meadows. Common. Flowers June-August.

Carex pauciflora — Few-flowered Sedge

Rhizomatous, 10-40 cm tall. Sheaths brownish, blades of lower sheaths obsolete, developed upwards, 0.5-1.8 mm broad, flat or more commonly margins rolled or channeled, lowermost bract scale-like, deciduous. Flower spikes solitary, androgynous, 0.5-1.5 cm long, 0.5-1.2 cm broad; pistillate scales almost equal to or shorter than perigynia, lance-attenuate, obtuse, pale brownish with thin, translucent margins and light median; perigynia long, reflexed, 6-7 mm long, 1-1.5 mm broad (rachilla not exserted), lance-linear, abruptly reflexed by a basal bend, stalkless, wingless, hairless, tapering to apex, several veined, greenish to straw-colored or brownish, apex thin, translucent to brownish; stigmas 3.

Mostly along coast from southeastern Alaska to Kodiak Island and Alaska Peninsula. Bogs. Common. Flowers June-August.

Carex pluriflora   Several-flowered Sedge

Loosely tufted, rhizomes purplish-black, 20-50 cm tall. Leaf sheaths purplish black to brown or tan, blades of lower sheath obsolete, elongating upwards, 1.2-4 mm broad, flat with rolled margins. Inflorescence 3-7.5 cm long, nodding (at least lower spikes), spikes 2-3, terminal staminate, lateral pistillate, with usually 10 or more fertile scales; pistillate scales longer than perigynia, ovate to elliptic, acute to tipped with a sharp rigid point, pale-green to purplish black with lighter median; perigynia 3.5-4.5 mm long, 1.7-2.3 mm broad, ovate, ascending, wingless, hairless, almost stalked, veined on both sides, greenish to brown, beakless, apex ending abruptly; stigmas 3.

Coasts and islands of southern Alaska and less commonly some distance inland in south-central Alaska. Bogs and wet meadows. Common. Flowers July-August. Includes *C. rariflora*.

Carex sitchensis   Sitka Sedge

More or less tufted, 40-100 cm tall from short, scaly, brown-purplish rhizome. Sheaths reddish brown, blades of lower sheaths obsolete, developed upwards, 2.5-9 mm broad, flat with rolled margins; lowermost bract 15-45 cm long or more, sheathless, blade leaf-like. Inflorescence 7-36 cm long, erect or more or less nodding, spikes 4-8, terminal 1-4 staminate, lower 2-5 pistillate (or some androgynous), separated by 10-150 mm, with erect to spreading or arched peduncles 0.6-8 cm long; pistillate scales longer than perigynia, lance-shaped to lance-attenuate, acute to attenuate, purplish brown to brownish, margins often thin and translucent; median straw-colored; perigynia 2.2-3 mm long, 1.2-1.7 mm broad, ovate to oval, appressed-ascending, wingless, hairless, veinless or nearly so, almost stipitate, greenish to straw-colored, beak 0.1-0.3 mm long, truncate; stigmas 2.

Along coast and on islands from southeastern Alaska west to Kenai Peninsula, rarely some distance into Interior Alaska. Salt marshes, meadows, and bogs. Common. Flowers June-August. Important wildlife food in wetlands.

Many *Carex* species are used in basket weaving. Stems used as bedding and mats for sitting in the sweat house.

Carex bigelowii (Bigelow Sedge) Rhizomatous, 10-60 cm tall; leaf sheaths brown to chestnut or purplish brown, blades of lower sheaths short to elongate, elongating upwards, 1-4 mm broad, flat or less commonly channeled below, margins rolled under; lower most bract 0.4-5 cm long, blades linear-awlshaped to leaf-like, sheathless; inflorescence 1-7.5 cm long, erect; spikes 2-5, terminal staminate, lateral pistillate or upper androgynous; pistillate scales almost equal to perigynia, elliptic to oblong or oblanceolate, acute to obtuse, purplish to brownish black with thin transparent margins and lighter midrib; perigynia 1.7-3.5 mm long, 1.2-2 mm broad, spreading-ascending, elliptic, hairless, short-stalked, veinless, greenish or straw-colored to purplish black, beak 0.1-0.3 mm long, entire apically; stigmas 2. Bogs, tundra, open woods, lake shores and meadows in most of Alaska.

Carex bitartita (Two-parted Sedge) Caespitose (often shortly rhizomatous), 10-40 cm tall; leaf sheaths brownish to straw-colored, blades of lower sheaths short or lacking, developed upwards, 1-3 mm broad, flat or channeled; lowermost bract scale-like, bladeless or blade 0.2-2 cm long and awl-shaped; inflorescence 1-3.5 cm long, erect; spikes 2-5; pistillate scales shorter than to equaling perigynia, brownish, ovate to elliptic, thin translucent margins, paler midrib; perigynia 2-4 mm long, 1-1.5 mm broad, lance-shaped to elliptic, ascending, veined on both faces, wingless, stalked, brownish or greenish, tapering to more or less flattened beak, beak 0.2-0.6 mm long, obliquely cleft; stigmas 2. Muskegs, stream banks, shallow ponds, wet meadows and open woods in most of Alaska. Flowers July-August. Also known as **C. lachenalii**.

Carex circinata (Coiled Sedge) Densely tufted, caespitose, 5-25 cm tall; leaves stiff, hair-like, curved or curled, sheaths light brown, blades of lower sheaths short, elongating upwards, less than 1 mm broad, margins rolled under, usually curved; spikes solitary 1.5-2.7 cm long, 0.3-0.5 cm broad, male and female flowers on same inflorescence; pistillate scales from shorter to longer than perigynia, lance-like to elliptic, obtuse to acute or acuminate, brownish with thin translucent margins and straw-colored mid vein; perigynia 4.5-6 mm long, 1-1.5 mm broad, erect-ascending, narrowly lance-like, tapering at both ends, veined dorsally and ventrally, short-stalked, wingless, hairless on both sides, minute teeth in margins, straw-colored, tapering to a slender reddish beak 1-2 mm long, white, thin transparent apically; stigmas 3. Ridge tops, rock crevices and meadows in coastal Alaska from southeastern Alaska to Aleutians. Flowers June-July.

202

Carex enanderi (Goose-grass Sedge) Rhizomatous or stoloniferous, or both (and loosely caespitose), 13-40 cm tall; leaf sheaths brownish to chestnut brown, blades of lower sheaths short or lacking, elongating upwards, 1-3 mm broad, flat to channeled throughout; lowermost bract 1.5-7.5 cm long, erect or nearly so; spikes 2-5, terminal with staminate and pistillate flowers, lateral pistillate, appressed-ascending to erect, all clustered, almost stalkless or less commonly lower separated by 12-25 mm with pedicels 0.6-1.2 cm long; pistillate scales slightly shorter than perigynia, lance-shaped to obovate, rounded to obtuse or apex notched, purplish black with straw-colored to greenish median; perigynia 2.5-3.5 mm long, 1.2-1.6 mm broad, ovate to elliptic, spreading-ascending, wingless, hairless, stipe 0.3-0.6 mm long, many-veined on both faces, straw-colored or purple spotted or purplish apically, abruptly beaked, beak 0.2-0.4 mm long, ending abruptly; stigmas 2. Bogs, wet swampy places, coasts and islands of southern Alaska. Also known as *C. eleusinoides.*

Carex flava (Yellow Sedge) Caespitose, 10-50 cm tall or more; leaf sheaths straw-colored to greenish, blades of basal leaves short or lacking, elongating upwards, 2-4 mm broad, flat; lowermost bract commonly 3-10 cm long, short-sheathing, spreading, blade leaf-like; inflorescence 2.5-6 cm long, erect; spikes 2-6, terminal staminate, lateral pistillate, upper commonly clustered, lower borne on peduncles 5-15 mm long or more; pistillate scales much shorter than perigynia, lance-shaped, acute to obtuse or with sharp rigid point, brownish, more or less thin and translucent, midrib greenish; perigynia 4-6 mm long, 1.5-2.5 mm broad, body lance-shape to oblanceolate or obovate, spreading or reflexed, wingless, hairless, stalkless, prominently veined, yellowish green or greenish, beak 2-3 mm long, obscurely double toothed; stigmas 3. Moist sites in southeastern and south-central Alaska.

Carex laeviculmis (Smooth-stem Sedge) Densely tufted, caespitose, 15-90 cm tall; leaf sheaths straw-colored to light brownish, blades of lower sheaths short or lacking, developed upwards, 1-2 mm broad, flat; inflorescence 1.5-6.5 cm long, erect; spikes 3-8, stalkless; pistillate scales shorter than perigynia, oval to ovate, obtuse to acute yellowish brown with broad, thin, transparent margins and greenish to brownish median; perigynia 2.2-3.5 mm long, 1-1.5 mm broad, lance-shaped to ovate, ascending, wingless, hairless, short stalked, veined on both faces, tapering to flattened minutely toothed beak, beak 0.2-0.8 mm long, obliquely cleft; stigmas 2. Gravelly slopes, bogs and wet woods near coasts and on islands of SE and southern Alaska. Flowers June-July.

203

Carex livida (Livid Sedge) Rhizomatous, seldom somewhat caespitose, solitary or loosely tufted, 10-50 cm tall, leaf sheaths light brownish, blades of basal sheaths well developed, 1-3 mm broad, channeled; inflorescence 2-8 cm long, erect or nearly so; spikes 2-4, terminal staminate, lateral pistillate; pistillate scales from shorter than to longer than perigynia, lance-ovate to elliptic, obtuse to acute, brownish to purplish brown with thin, translucent margins and greenish to straw-colored median; perigynia 3-4.5 mm long, 1.5-2.5 mm broad, elliptic, ascending, wingless, hairless, stalked, veinless, greenish to whitish or straw-colored, beak about 0.1 mm long, ending abruptly; stigmas 3. Bogs in southeastern, southern and northwestern Alaska.

Carex macrocephala (Large-headed Sedge) Dioecious (rarely monoecious), rhizomatous, 10-40 cm tall; leaf sheaths brownish, sometimes obscured by old leaves, blades of basal sheaths short or lacking, elongating upwards, 4-10 mm broad, channeled; lowermost bract scale-like or to 6 cm long and blade leaf-like; inflorescence mostly 4-7 cm long and 1-4 cm broad, erect, pistillate inflorescence with several to many large spikes, all stalkless; pistillate scales shorter than perigynia, lance-shaped, acuminate to awned, chestnut brown with thin,

translucent margins and green mid vein or with broad straw-colored median; perigynia 10-12 mm long, 4-6 mm broad, lance-shaped, ascending, winged, minutely spiny and toothed, hairless, short-stalked, strongly veined on both sides, brownish, beak 4-6 mm long, minutely toothed, sharply doubly-toothed; stigmas 3. Sandy seashores from southeastern Alaska to Alaska Peninsula. Flowers July-August.

Carex nigricans (Black Alpine Sedge) Tufted from stout creeping rhizomes, 5-40 cm tall; leaf sheaths brownish to straw-colored, blades of lowermost sheaths short or lacking, elongating upwards, 1-2.5 mm broad, flat or channeled, lowermost bract scale-like; spikes solitary, 0.8-1.7 cm long, 0.5-0.9 cm broad, with both sexes; pistillate scales deciduous, about equaling perigynia, lance-shaped to ovate, obtuse to acute or attenuate, brown with lighter midvein; perigynia 3.5-4.8 mm long, 1-1.8 mm broad, lance-shaped, ascending to spreading, veinless, wingless, hairless, with stipes 0.6-1.2 mm long, brownish, beak 0.3-0.6 mm long, oblique apically, thin, translucent; stigmas 3. Alpine meadows and mountain slopes on coasts and islands from southeastern Alaska west to Aleutians. Flowers in July.

CYPERACEAE

Carex phaeocephala (Darkhead Sedge) Caespitose, 8-25 cm tall; leaf sheaths reddish-brown, blades of lower leaves lacking or short, developed upwards, 1-3 mm broad, flat or channeled to margins rolled; lowermost bract 0.5-2 cm long, scale-like or blade awl-shaped; inflorescence 1.5-3.5 cm long, 0.8-1.7 cm broad, erect or nearly so; spikes 2-6, with pistillate and staminate flowers in same spike, all clustered or lower separated by 2-11 cm, all stalkless or lower peduncles to 3 mm long; pistillate scales almost equal to, more or less obscuring perigynia, lance-shaped, acute to obtuse, brown with white, thin, translucent margins and lighter median; perigynia 4-6 mm long, 1.3-2.5 mm broad, lance-shaped, appressed-ascending, winged, hairless, except for minutely toothed margins, almost stalked, lightly veined dorsally, straw-colored to brownish, tapering to flattened beak, beak 0.8-1.2 mm long, obliquely cleft, apex thin, translucent; stigmas 2. Wet meadows in SE and south-central Alaska, alpine.

Carex phyllomanica (Stellate Sedge) Caespitose, 14-65 cm tall; leaf sheaths straw-colored to brownish, blades of lower sheaths obsolete, elongating upwards, 1-3.5 mm broad, flat or channeled; lowermost bract 0.2-1.5 cm long, scale-like or blade linear-awlshaped; inflorescence 1.5-2.5 cm long; spikes 2-4(5), stalkless, ascending-spreading; pistillate scales shorter than perigynia, ovate, acute to obtuse, greenish or brownish with thin, translucent margins and lighter median; perigynia 3-4.5 mm long, 1.2-1.7 mm broad, lance-shaped, spreading, wingless (barely sharp-edged), hairless, stalkless, veined on both sides, straw-colored to brownish, tapering to a flattened minutely toothed beak, beak 1-1.5 mm long, bidentate; stigmas 2. Bogs and fens on coasts and islands of southeastern and southern Alaska; less common in Interior. Also known as **C. echinata**.

Carex praticola (Meadow Sedge) Caespitose to 70 cm tall, leaf sheaths brownish to straw-colored, blades of lower sheaths short, elongating upwards, 2-5 mm broad, flat; lowermost bract 0.3-5.5 cm long, blade (when present) linear-awlshaped; inflorescence 2-7 cm long, erect to nodding; spikes 3-10, male and female, tapering to slender staminate base, stalkless, clustered; pistillate scales shorter than to almost equal to perigynia, lance-shaped, acuminate to acute, yellowish brown except for green median, thin, translucent margins; perigynia 3.5-5.2 mm long, 1.3-1.8 mm broad, lance-shaped to lance-ovate, appressed-ascending, wing-margined and minutely toothed, hairless, almost stalked, veined on both sides , green to brownish, tapering to slender flattened beak, beak 1-2 mm long, minutely toothed, obliquely cleft; stigmas 2. Open woods, bogs, bluffs, floodplains, meadows from SE to south-central Alaska. Includes **C. aenea**.

205

Carex pyrenaica (Pyrenean Sedge) Caespitose, 10-30 cm tall; leaf sheaths brownish to chestnut, blades of lower sheaths obsolete, developed upwards, 0.5-1.5 mm broad, channeled; lowermost bract scale-like; spikes solitary, male and female, 0.8-2.1 cm long, 0.4-0.7 cm broad, erect; pistillate scales shorter than to almost equal to perigynia, lance-shaped, attenuate to acute, brown to chestnut with thin, translucent margins and lighter median; perigynia 2.8-4.1 mm long, 0.9-1.3 mm broad, lance-shaped, ascending, wingless, veinless, stalked, (stalk 0.4-0.8 mm long), brownish to straw-colored, tapering to an elongate hairless beak, beak 0.2-0.6 mm long, oblique, thin, translucent apically; stigmas 3. Alpine meadows from SE Alaska west to Aleutians, central Alaska. Flowers June-July.

Carex saxatilis (Russet Sedge) Rhizomatous, 15-100 cm tall; leaf sheaths chestnut brown to purplish or reddish, blades of lower leaves obsolete, developed upwards, 2-4.5(6.4) mm broad, flat with rolled margins or channeled below; lowermost bract 2-21 cm long, sheathless or rarely with a cylindrical sheath 1-3 mm long or more, blade leaf-like; inflorescence 5-16 cm long, erect to more or less nodding; spikes 2-4, terminal 1-2(3) staminate, lower 1-3 pistillate; pistillate scales shorter than perigynia, lance-shaped to ovate, acute to acuminate or almost obtuse, purple brown or purple black, margins narrowly to broadly thin, translucent, midrib lighter and commonly with a broadly thin, translucent apex; perigynia 3.5-5.7 mm long, 1.5-2.5 mm broad, ovate to lance-shaped or lance-oblong, ascending to spreading-ascending, wingless, hairless, almost stalkless, veinless or nearly so, membranous, purplish brown to purplish black or yellow to greenish and often purplish mottled, shining at maturity, beak 0.4-0.7 mm long, ending abruptly or with a shallow apical notch or shortly bidentate; stigmas 2 (3). Ponds, bogs, stream banks, floodplains and tundra throughout Alaska.

Carex scirpoidea (Single-spike Sedge) Rhizomatous, dioecious, 10-40 cm tall; leaf sheaths purplish brown to purplish black or purplish, blade of lower sheaths lacking, developed upwards, 1-3 mm broad, flat or channeled; lowermost bract 0.5-1.5 cm long or more, sheathless, blade awl-shaped to leaf-like; spikes solitary, erect; pistillate scales shorter than to longer than perigynia, ovate to lance-shaped, obtuse to acute, purplish brown, more or less thin, translucent fringed margins and light to brown median; staminate scales brown, fringed; perigynia (2.4)3-4 mm long, (0.9)1.1-1.7 mm broad, obovate to elliptic, appressed-ascending, veinless, wingless, more or less densely hairy, almost stalked, purplish brown to straw-colored, beak 0.2-0.5 mm long, ending abruptly or with a shallow apical notch, thin, translucent apically; stigmas 3. Bogs, rock outcrops, woods and tundra mid to high elevation in most of Alaska except southwestern portion. Flowers June-August.

Carex stipata (Sawbeak Sedge) Caespitose, 25-90 cm long or more; leaf sheaths flaccid, brown to straw-colored, blades of basal sheaths obsolete, developed upwards, mostly 3-12 mm broad, flat; inflorescence 1.5-5.5 cm long, erect; spikes several to many in simple or compound clusters, with both male and female, stalkless, ascending; pistillate scales shorter than perigynia, lance-shaped to ovate, acute to attenuate or shortly awned, brownish to yellowish brown and more or less thin, translucent, median straw-colored to greenish; finely toothed perigynia 4-5 mm long, 1-2 mm broad, lance-attenuate, ascending, wingless, hairless, short-stalked, veined on both sides, brownish to straw-colored, tapering to a slender minutely toothed beak, 2-2.5 mm long, beak bidentate; stigmas 2. Streamsides, wet meadows and bogs, low elevation in coastal southeastern to south-central Alaska. Flowers July.

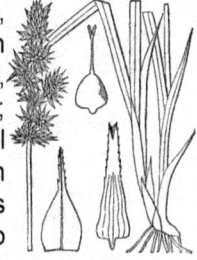

Carex stylosa (Variegated Sedge) Densely tufted, caespitose, 10-50 cm tall; leaves branched near base, leaf sheaths brown, blade of lower sheaths short or lacking, developed upwards, 0.8-3 mm broad, flat with rolled margins or channeled below; lowermost bract 1.2-4 cm long, sheathless, blade leaf-like; inflorescence 3-6 cm long, erect or more or less nodding; spikes 2-4, terminal staminate, lateral pistillate, with usually 10-more fertile scales; pistillate scales from shorter than to longer than perigynia, ovate, obtuse to acute, purplish black with thin, translucent margins and lighter median; perigynia 2.5-3.5 mm long, 1-1.5 mm broad, ovate to elliptic, ascending, wingless, hairless, short-stalked, veinless, straw-colored to brownish or tinged purplish, beak 0.1-0.3 mm long, ending abruptly; stigmas 3. Bogs low to mid elevation from coastal southern Alaska into Interior.

Eleocharis palustris   Creeping Spike-rush

Stems closely clustered, striate, erect from brownish black rhizome, 10-150 cm tall. Leaves basal, leaf sheaths gold or reddish below, somewhat flaring upwards, bladeless. Spikes solitary, round or nearly so, subtended by 2 or 3 bracts encircling half the stem, scales reddish brown or purplish brown with clear margins. Achene light yellow to brown, smooth, tubercle conic-triangular, flattened, longer than broad, much smaller than or as long as achene, 4 bristles.

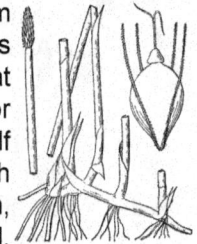

Much of Alaska south of Brooks Range. Bogs, river and stream banks and pond margins. Common. Flowers late May-July. Includes *E. uniglumis*.

Eleocharis kamtschatica (Kamchatka Spikerush)
Caespitose to erect, perennial, more or less rhizomatous, 10-40 cm tall, stems 0.8-1.5 mm in diameter; leaf sheaths purplish below, slightly flaring upwards, bladeless; spikelets solitary, 6-15 mm long m long, 0.3-0.7 mm broad, flattened, subtended by a single bract almost or encircling stem; scales 3.5-5.5 mm long, purplish black to purplish brown with thin transparent margins and reddish midvein; achenes 1.3-1.8 mm long, marked in longitudinal rows, tubercle slightly smaller than achene, about as broad, cap-like; perianth bristles 4-6. Tide flats and saline meadows in coastal southeastern, south-central and west-central Alaska at low elevation.

Eleocharis quinqueflora (Few-flowered Spikerush)
Caespitose and rhizomatous, mostly 6-35 cm tall, stems 0.5-1 mm in diameter; leaf sheaths brownish to straw-colored below, cylindrical, not flaring upwards, bladeless; spikes solitary, 4-9 mm long, 2-4 mm broad, flattened, subtended by two bracts encircling about half stem and about half as long as spike or more; scales mostly 3-5 mm long, brown to purplish brown with thin transparent margins and lighter midvein; achenes 1.8-2.2 mm long, straw-colored, with leaf veins in a minutely network, tubercle not or poorly developed; perianth bristles 4-6. Marshes and bogs in Montana Creek area near Juneau and interior central to eastern Alaska.

KEY TO SPECIES OF ***ERIOPHORUM:***

1a. Spikes 2-several (rarely some solitary); inflorescence subtended by 1-more leafy bracts. **(2)**

1b. Spikes solitary, not subtended by leafy bracts. **(4)**

    2a. Leaves 1-2 mm broad, channeled throughout; peduncles minutely short-hairy overall; plants uncommon in southeastern, south-central, and northwestern Alaska.
                                        E. gracile

    2b. Leaves commonly (1.5)2-4 mm broad or more, at least some, flat below middle; peduncles various. **(3)**

    3a. Anthers 1-1.3 mm long; midrib of scale prominent at tip; peduncles minutely hairy throughout; plants uncommon.
                                  E. viridicarinatum

    3b. Anthers 2-5 mm long; midrib of scale not prominent at tip; peduncles hairless or merely with sharp forward pointing teeth marginally; plants common and widespread.   ***E. angustifolium***

4a. Anthers 0.5-1.5 mm long; spikes mostly less than 2.5 cm long; bristles white.   ***E. scheuchzeri***

4b. Anthers 1.5-3(3.2) mm long; spikes mostly over 2.5 cm long; bristles cinnamon brown or white.   ***E. chamissonis***

Eriophorum angustifolium
Narrow-leaved Cotton-grass

Creeping rhizomes covered at base with brown or purple persistent sheaths, culms smooth, 30-70 cm tall, almost circular in transverse section. Leaves usually flat below middle, triangular, channeled or folded above middle, stem leaves commonly 2 or more, blades 2-6 mm broad, involucral leaves 2-3, purplish basally. Flower spikes 2-10, at least some usually drooping, peduncles hairless or with small marginal teeth, in flower ovoid and 1-2 cm long, in fruit 2-4 cm long; scales grayish, purple-green or brown, ovate to lance-shaped, appressed-ascending, midvein not reaching apex, somewhat acute; anthers mostly 2.5-5 mm long. Achenes 2.5-3.5 mm long, bristles numerous, white to tawny, up to 3 cm long.

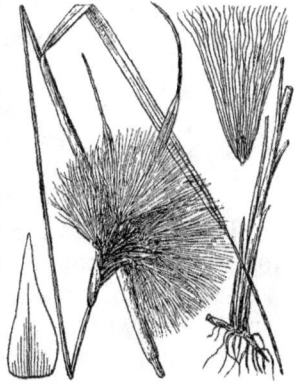

Throughout Alaska. Bogs, meadows, alpine meadows and open areas from low to high elevation. Common. Flowers late June-August. Stem base eaten raw; upper part of underground stem immersed in boiling water to remove black covering and eaten. Also known as *E. polystachion*.

Eriophorum chamissonis
Russett Cotton-grass

From spreading, creeping rhizome, culms stout, mostly 20-70 cm tall, covered basally with brown to purplish brown persistent sheaths. Leaves channeled throughout, threadlike, 0.4-1.3 mm broad, stem leaves usually 1-3, placed below middle of stem (rarely 1 above), sheaths not or only somewhat expanded upwards; involucral leaves lacking. Spikes solitary, erect, in flower oblong-cylindrical, 1.5-2 cm long, in fruit obovoid to almost spherical, 18-40 dm long; lowermost scale (spathe) 4-10 dm long, sterile scales mostly less than 7, scales blackish to grayish, ovate-lance-shaped, appressed-ascending, attenuate, midvein not reaching apex; anthers 0.5-1.5 mm long. Achenes 1.7-2.3 mm long, smooth along margin near tip, bristles numerous, cinnamon to white.

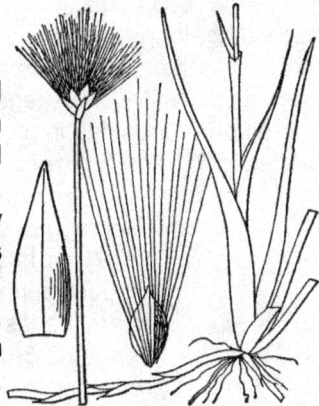

Most of Alaska. Bogs, lake shores, bogs, wet meadows and stream banks. Flowers June-August. Also known as *E. russeolum*.

Eriophorum gracile   Slender Cotton-grass

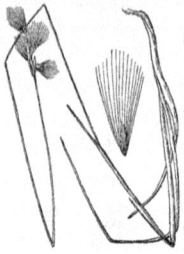

Rhizomatous, culms 20-60 cm tall, slender, commonly lacking young basal leaves at flowering. Linear leaves channeled throughout, stem leaves 2-3, blades not over 2 mm wide, purplish to brownish basally. Spikes 2-8, at least some drooping, pubescent peduncles, in flower ovoid and 5-10 dm long, in fruit 15-50 dm long; scales grayish or blackish, lance-ovate to lance-shaped, appressed-ascending, midvein not reaching apex, obtuse except uppermost; anthers 1-2 mm long. Achenes 2.5 mm long, linear-oblong; bristles numerous, white, 5-25 mm long.

South-central and less commonly in northwestern and southeastern Alaska. Peaty soil in bogs and marshes. Occasional. Flowers July-August.

Eriophorum scheuchzeri (White Cotton-grass) Rhizomatous, culms 2-6 dm tall, clothed basally with brownish to reddish partitioned, knobby persistent sheaths; leaves channeled throughout, 0.5-1.3 mm broad; stem leaves 1(2) below middle of stem, sheath slightly expanded upwards; involucre leaves lacking; spikes solitary, erect, in flower almost globe-like to oblong-cylindrical, 0.8-2 cm long, in fruit obovoid to almost globe-like, 1.8-4 cm; lowermost scale (spathe) 0.4-1 cm long; sterile scales mostly less than 7; scales blackish to grayish, ovate-lance-shaped, appressed-ascending, long-tapering, midvein not reaching apex; anthers 0.5-1.5 mm long; achenes 1.7-2.3 mm long, not ciliate-minutely toothed; bristles numerous, white. Muskegs, meadows, river banks, lake and pond shores and roadsides in most of Alaska. Flowers June-July. Closely related to *E. chamissonis*.

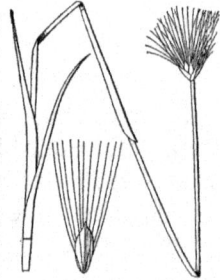

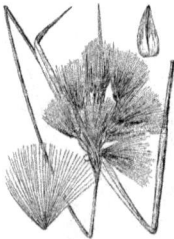

Eriophorum viridicarinatum (Green-keeled Cotton-grass) Similar to *E. angustifolium*. Rhizomatous, 20-70 cm tall, covered basally with brownish partitioned, knobby persistent sheaths; leaves flat except at very tip; stem leaves 2-more, blades 2-6 mm broad; involucre leaves 2-3(4), green to brownish basally; spikes 2-10 or more, at least some drooping, peduncles minutely hairy throughout, in flower obovoid and 15-30 mm long; scales grayish to greenish, lance-shaped to long tapering, appressed-ascending, conspicuous midvein extending to tip; anthers 1-1.3 mm long; achenes 3-4 mm long; bristles numerous, creamy white to tawny or slightly yellowish. Meadows and bogs in southeastern and south-central Alaska.

Rhynchospora alba   White Beak-rush

Tufted, from short or mostly absent rhizomes, 15-50 cm tall. Leaf sheaths straw-colored, more or less persistent, blades 0.3-1 mm broad, linear, lowermost bract 0.5-6 cm long, sheathless or with a closed cylindrical sheath, 1-15 mm long. Flower spikelets in 1-4 dense head-like clusters, pale brown to nearly white; scales ovate to lance-ovate, whitish; bristles with backward facing barbs. Achenes brownish-green with stiff bristles, 1.5-2.5 mm long, with a sharp tubercle about one-third to one-half as long as achene.

Southeastern and southwestern Alaska. Bogs, fens, wet peaty or sandy soil, low to mid elevation. Uncommon. Flowers July-August.

Scirpus caespitosus   Tufted Clubrush

Densely tufted though sometimes with rhizomes connecting tufts or absent, culms circular in transverse section or nearly so, 10-40 cm long, 1.5 mm in diameter. Leaf sheaths straw-colored to brownish, blades lacking or upper ones with blades to 8 mm long, leaves all near base of culm. Flower spikelets solitary, 3.5-6 mm long, terminal, subtended by a solitary scale about as long as spikelet; scales ovate, yellowish brown, lower 2-3 bluntly awn tipped; bristles 6, smooth, twice as long as achene. Achenes compressed, 1.5-2 mm long, triangular.

Most of Alaska (except central valleys). Wet areas, bogs, ponds and meadows. Common. Flowers July-August. Can turn bright yellow or orange in fall. Also known as *Trichophorum cespitosum*.

Scirpus microcarpus   Small-flowered Bulrush

Culms triangular stout and leafy to 15 dm tall, from a rhizomatous woody base, 4-13 mm in diameter near base. Leaf sheaths purplish to brownish tinged or straw-colored to greenish, with partitioned knobs, blades all well developed, 7-20 mm broad, leaves distributed along most of stem. Flower spikelets numerous, 3-6 mm long, borne in an open paniculate cluster, inflorescence subtended by 2-3 or more leaf-like bracts, scales ovate, tinged purplish black to greenish; bristles 4, about as long as achene, with backward facing barbs. Achenes 1-1.5 mm long.

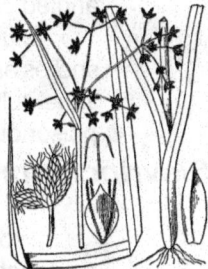

Southeastern Alaska north throughout Alaska. Bogs, pond margins, meadows, arctic and alpine tundra and woods, low to mid elevation. Common. Flowers July-August. Also known as *S. sylvaticus* and *S. rubrotinctus*. Used to make baskets and to line steaming pots.

Scirpus maritimus (Seacoast Bulrush) Rhizomatous, culms triangular, up to 15 dm tall, 5-20 mm in basal diameter; leaf sheath brownish, partitioned with small knobs; blades well-developed, leaves distributed along stem; spikelets three to 20 terminal, densely clustered at tip of stem, nestled in 3 or more leafy bracts,12-20 mm long by 6-10 mm wide; scales ovate, brownish to brown, 2-toothed and awned at tip; bristles 1-3, much shorter than achene; achenes 2.5-4 mm long. Low elevation wet beaches and tide flats south-central and southeastern Alaska where rare. Also known as ***S. paludosus, Schoenoplectus maritimus*** and ***Bolboschoenus maritimus***.

Scirpus tabernaemontanii (Great Bulrush) Coarsely rhizomatous, culms circular in transverse section, 5-15 dm tall or more, 6-20 mm in basal diameter; leaf sheaths brown to straw-colored or mottled brownish, usually all bladeless; leaves all near culm base; spikelets several to many, 5-12 mm long in compound compact to open branched raceme cluster subtended by 1 (rarely more) reduced leaf-like bract to 7 cm long; scales ovate, brown to purplish brown, shortly awned; achenes 1.5-2.5 mm long. Variable species. Tide flats, bogs, ponds and lakes in southeastern third (including Panhandle) of Alaska. Flowers July-August. Also known as ***S. validius, S. lacustris*** and ***Schoenoplectus tabernaemontani***. Used for baskets, mats and temporary shelters.

IRIDACEAE
IRIS FAMILY

Iris setosa   Wild Flag

Perennial 3-7 dm tall from short, thick rootstock with remains of old leaves. Leaves mainly basal, sword-shaped to linear-lance-shaped. Flowers large, showy, sepals contracted into short claws, petals small with narrow blade. Sepals bluish to purplish, dark veined. Fruit egg-shaped, bluntly angled capsules.

Much of Alaska south of Brooks Range. Meadows, shores and tide flats, low to mid elevation. Common. Flowers June-July. Rhizome contains poison irisin; seeds may be poisonous as well. Iris is Latin for rainbow.

Sisyrinchium angustifolium   Blue-eyed Grass

Tufted herb 1-4 dm tall with simple, broadly to narrowly winged stem. Leaves mostly basal, grasslike, bracts greenish or purplish tinged. Flower umbels 3-6-flowered, subtended by paired, erect, green bracts. Flower color blue to violet with yellow eye. Fruit spherical capsule on short stalk, seeds black.

Southeastern Alaska west to Aleutians. Meadows, moist places along beaches, low to mid elevation. Occasional. Flowers May-July. Also known as ***S. montanum;*** includes ***S. littorale***.

JUNCACEAE
RUSH FAMILY

KEY TO SPECIES OF *JUNCUS:*

1a. Plants annual; leaves less than 1 mm broad; rhizomes lacking.   *J . bufonius*

1b. Plants perennial; leaves often more than 1 mm broad; rhizomes often present. **(2)**

2a. Lowermost bract of inflorescence circular in transverse section and erect, seemingly representing a continuation of stem axis, inflorescence thus apparently lateral. **(3)**

2b. Lowermost bract of inflorescence flattened or variously channeled, not circular in transverse section, erect to spreading, inflorescence obviously terminal. **(6)**

    3a. Lowermost bract 1-3 cm long; flowers 1-4 per stem.   *J. drummondii*

    3b. Lowermost bract usually over 5 cm long; flowers (4)5-many per stem. **(4)**

    4a. Perianth 4-6.5 mm long; stems 0.9-2.7 mm broad below inflorescence.   *J. arcticus*

    4b. Perianth 2-3.5(4) mm long; stems various, but if more than 0.9 mm broad than capsules sharply angled apically. **(5)**

    5a. Involucral bract rarely half as long as stem; stems 0.9-2 mm broad below inflorescence; capsules sharply angled above.   *J. effusus*

    5b. Involucral bracts commonly more than half as long as stem; stems 0.5-0.9 mm broad below inflorescence.   *J. filiformis*

6a. Plants with a scape, leaves all basal; heads solitary, 1-3(5)-flowered. **(7)**

6b. Plants with leafy stems, leaves not all basal; heads more than one or with more than 5 flowers. **(8)**

    7a. Involucral bract almost equal to head; capsule acute to obtuse apically; heads commonly 3-flowered.   *J. triglumis*

    7b. Involucral bract usually surpassing head; capsule with a rounded shallowly notched apex; heads commonly 2-flowered.   *J. biglumis*

8a. Leaf blades laterally flattened with one edge toward stem, leaves equally spaced; plants of coasts and islands of southern Alaska.   *J. ensifolius*

8b. Leaf blades circular in transverse section or dorsoventrally compressed, leaves not equally spaced; plants of various distribution. **(9)**

9a. Heads 4-more or flowers loosely clustered; perianth 2.5-3.5 mm long (longer in some *J. supiniformis*). **(10)**

9b. Heads 1-3; perianth 3.5-5 mm long or more (shorter in some *J. nodosus*). **(14)**

    10a. Heads spherical; capsules attenuate apically.   *J. nodosus*

    10b. Heads hemispherical; capsules abruptly short beaked apically, not attenuate. **(11)**

    11a. Flowers loosely arranged, not densely clustered in tight heads; perianth surpassing capsule.   *J. tenuis*

    11b. Flowers densely aggregated into definite heads; perianth shorter than capsule. **(12)**

    12a. Stems often decumbent and rooting at nodes, often with leaves proliferating from heads; plants of coastal southeastern and south-central Alaska.   *J. supiniformis*

    12b. Stems erect or ascending, not proliferating leaves from heads; plants of various distribution. **(13)**

    13a. Outer perianth segments acute, inner ones obtuse; plants of broad distribution.   *J. alpinus*

    13b. Outer perianth segments and inner ones both acute; plants of southern Alaska.   *J. articulatus*

14a. Capsules 5-8 mm long; seeds 2.5-3.5 mm long or more, appendaged at both ends; bracts, perianth, and capsule all brown. ***J. castaneus***
14b. Capsules less than 5 mm long; seeds less than 2 mm long, not appendaged; bracts, perianth, and capsules various but sometimes all brown. **(15)**
15a. Heads commonly 2-3, spherical; capsules attenuate apically. ***J. nodosus***
15b. Heads commonly solitary (rarely 2 or more), hemispherical; capsules abruptly beaked apically. **(16)**
16a. Heads 1-3(5)-flowered; leaves less than 1 mm broad. ***J. stygius***
16b. Heads more than 5-flowered; leaves 1-3 mm broad (at least some). ***J. falcatus***

Juncus arcticus   Arctic Rush

Perennial from long-creeping, horizontal rhizomes, culms erect, arising along rhizome in comb-like fashion, 10-10 dm tall, circular in transverse section or somewhat flattened. Leaves basal, commonly reduced to bladeless yellowish-brown sheaths. Inflorescence lateral, 3-many, involucral bract erect, circular in transverse section, flowers congested or borne in loose cluster, perianth 4-6.5 mm long, brownish to dark brown, outer ones differing from inner in size and often texture, stamens 6, anthers as long as filaments or much longer. Fruit a more or less ovoid capsule, seeds minutely appendaged.

Most of Alaska. Sandy shores, wet areas and salt marshes, low to mid elevation. Common. Flowers June-August. Considered part of the Juncus arcticus-balticus complex.

Juncus bufonius   Toad Rush

Annual, stems cespitose, solitary or tufted from fibrous roots, culms .5-4 dm tall, channeled, simple or more commonly branched. Leaves 1-several per culm, both basal and stem leaves present, blades channeled, narrowly linear-awlshaped. Inflorescence terminal, commonly much branched, flowers solitary, subtended by thin, dry translucent bracts, primary bract shorter than inflorescence, perianth greenish or thin, dry, translucent, 2.5-8.5 mm long, segments lance-acuminate, with greenish midribs and broad thin translucent margins; stamens usually 6, anthers shorter than filaments. Capsules oblong-ellipsoid, rounded at tip, chestnut brown, usually shorter than perianth; seeds not appendaged.

Southeastern two thirds of Alaska. Pond and stream margins, tide flats, wet places in general, low to mid elevation. Common. Flowers July-September. A widespread, variable complex of species apparently consisting of native and introduced varieties.

Juncus effusus   Bog Rush

Perennial from short-branched rhizomes forming distinct, often large clumps, culms erect, solitary or few, 4-13 dm tall, circular in transverse section. Leaves basal, reduced to sheaths or uppermost with bristle-like blades. Inflorescence apparently lateral, compact to open, many-flowered; perianth greenish, 2-3.2 mm long; segments lance-attenuate, with greenish or straw-colored broad midribs and narrow translucent margins; stamens 3, anthers almost equal to filaments. Capsules greenish tan or darker, ovoid to obovoid, 1.8-2.2 mm long, obtuse apically, usually shorter than perianth; seeds amber, not appendaged.

Bogs and pond margins. Coast and islands of southeastern Alaska. Common. Flowers July-August.

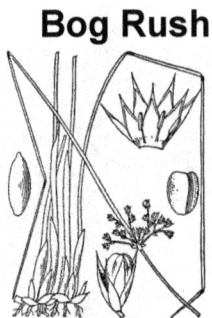

Juncus ensifolius   Dagger-leaved Rush

Perennial from creeping rhizomes 2-3 mm diam, culms erect, solitary or few, 2-6 dm tall, flattened. Leaves basal, 1-3, sword-shaped, blades well developed. Inflorescence terminal, lowermost bract 0.5-10 cm long, shorter to longer than inflorescence; heads 1-7 or inflorescence compound and with numerous small heads, each few to many-flowered; perianth brown, 2.8-3.7 mm long, segments lance-oblong to -attenuate, acute to attenuate; stamens 3, anthers shorter than filaments. Capsules oblong, almost equal to perianth or shorter, obtuse apically; seeds elliptic to obovate, 0.4-1 mm, occasionally tailed.

Southeastern Alaska west to Aleutians. Pond margins, bogs, woods and alpine tundra, low to subalpine elevation. Common. Flowers July-August.

Juncus alpinus (Alpine Rush) Perennial from short, creeping rhizomes, culms tufted, 15-50 cm tall, circular in transverse section or nearly so, leaves 1-3 per stem, blades circular in transverse section; inflorescence terminal; heads 5-many, 3-10-flowered, flowers stalkless or nearly so; perianth light brown to straw-colored, 2-2.5 mm long, inner segments short and rounded, outer acute; stamens 6, anthers shorter than filaments; capsule ovoid, almost equal to perianth; seeds minutely appendaged at each end. Sandy banks and shores and moist sites near lakes, streams and ponds in most of Alaska. Flowers in July.

215

Juncus articulatus (Jointed Rush) Perennial from short rhizomes, culms tufted, 10-50 cm tall, circular in transverse section or nearly so; leaves 1-3 per stem, blades circular in transverse section; inflorescence terminal, branches spreading; heads 5-many, 6-12-flowered; perianth brown to straw-colored, 2.5-3 mm long, segments acute to tapering to apex; stamens 6, anthers shorter than filaments; capsule ovoid, almost equal to or longer than perianth; seeds minutely appendaged at each end. Pond and lake margins in southeastern Alaska where evidently rare.

Juncus biglumis (Two-flowered Rush) Perennial, tufted, from short rhizomes, culms 2-30 cm tall, channeled along one side; leaves 1-4 per culm, all basal or nearly so, blades circular in transverse section; inflorescence terminal, commonly slightly surpassed by lowermost bract, stalkless; heads solitary, 2, 3 or 4-flowered; perianth brownish purple, 2.3-4.5 mm long, segments acute to obtuse; stamens 6, anthers shorter than filaments; capsule obovoid, with a rounded apical notch, 3-5.5 mm long, much longer than perianth; seeds minutely appendaged at each end. Sphagnum mats, pond and lake margins, stream shores and in shallow water in most of Alaska. Flowers in June.

Juncus castaneus (Chestnut Rush) Perennial from elongate rhizomes, culms solitary, 8-60 cm tall, circular in transverse section or nearly so, leaves 3-5 per culm, basal and on stem, blades channeled; inflorescence terminal, commonly surpassed by lowermost bract; heads 1-3 rarely 5, 2-7-flowered; perianth purplish brown to brown, 2.9-4.8 mm long, segments acute to acuminate; stamens 6, anthers shorter than filaments; capsule oblong-ellipsoid, 5-8 mm long, acute to obtuse and beaked apically, longer than perianth; seeds appendaged at each end. Moist meadows, bogs, woods and tundra. Northern southeastern north throughout most of Alaska. Flowers June-August.

Juncus drummondii (Drummond's Rush) Perennial from short rhizomes, culms tufted, 5-45 cm tall, circular in transverse section or nearly so, leaves all basal, reduced to sheaths or upper with bristle-like blades; inflorescence appears lateral; commonly equal to or surpassed by circular in transverse section lowermost bract, flowers solitary, usually 1-3 per inflorescence; perianth brownish, 4.8-6.5 mm long, segments thin and translucent, acute to acuminate; stamens 6, anthers almost equal to or longer than filaments; capsule oblong-ellipsoid, 4-7 mm long, truncate or with shallow apical notch, almost equal to or longer than perianth; seeds appendaged at each end. Moist meadows, open woods and alpine tundra from southeastern Alaska west to Aleutians, mostly at high elevation. Flowers in July.

216

Juncus falcatus (Sickle-leaved Rush) Perennial, from slender, creeping rhizomes, culms solitary or few, 10-30 cm tall, circular in transverse section or nearly so; leaves basal (few to several) and on stem (solitary), blades well developed; inflorescence terminal; heads solitary or sometimes 2-3, 5-many-flowered; perianth brown, 3.5-6 mm long, segments lance-shaped, attenuate or acute apically; stamens 6, anthers shorter to longer than filaments capsules, obovoid, 3.5-5 mm long, shorter than perianth; seeds not appendaged. Coastal marshes and tide flats on islands and coasts of southeastern and southwestern Alaska.

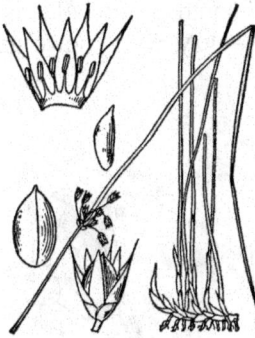

Juncus filiformis (Thread Rush) Perennial, rhizomatous, culms more or less tufted, 7-40 cm tall, circular in transverse section; leaves all basal, reduced to sheaths or upper with bristle-like blades; inflorescence appears lateral; flowers few to several in more or less open cluster; perianth greenish to brownish, (2.5)2.8-4.3 mm long, segments lance-attenuate with broad, thin, translucent margins; stamens 6, anthers shorter than filaments; capsules ovoid to obovoid, almost equal to perianth, obtuse apically; seeds not appendaged. Wet sites, meadows and woods low to subalpine elevation in much of southern Alaska south of 66th parallel (except Aleutians). Flowers June-July.

Juncus nodosus (Tuberous Rush) Perennial, rhizomatous, culms arising singly, 15-50 cm tall, circular in transverse section; leaves on stem 1-3 and basal (few), blades well developed (half rounded); inflorescence terminal, usually exceeded by lowermost bract; heads 3-many, many-flowered, perianth greenish or brownish, 3-4 mm long, segments lance-awlshaped; stamens 6, anthers almost equal to or shorter than filaments; capsule lance-ovoid, longer than perianth; seeds minutely sharp pointed at each end. Bogs in coastal southeastern and central interior Alaska.

Juncus stygius (Filiform Rush) Perennial, not rhizomatous, culms tufted, 10-25 cm tall, thread-like, circular in transverse section; leaves 1-3 per stem, blades thread-like, circular in transverse section; inflorescence terminal, commonly shorter than lower involucre bract, unbranched; heads 1 or 2(4), 1 to 4-flowered; perianth pale with red midribs and thin translucent margins, 3-4 mm long; stamens 6, anthers much shorter than filaments; capsule ovoid, longer than perianth; seeds appendaged at each end. Bogs and shallow pools in southeastern and south-central Alaska.

Juncus supiniformis (Spreading Rush) Perennial, stems generally tufted along slender rhizomes, 10-30 cm tall; leaves 2-4 per stem, blades half round; inflorescence terminal, usually surpassing lowermost involucre bract, commonly branched; heads 2-6, 3- to several-flowered; perianth pale brown to dark brown, 3-4.5 mm long, segments acute to acuminate; stamens 3 or 6, anthers shorter than filaments; capsules cylindrical, longer than perianth; seeds shortly appendaged at each end. Bogs and ponds on coast and islands of southeastern and south-central Alaska, low to mid elevation. Flowers June-July. Also known as ***J. oreganus***.

Juncus tenuis (Slender Rush) Perennial, tufted, 15-60 cm tall, culms circular in transverse section or nearly so, blades flattened; inflorescence terminal, commonly exceeded by lowermost involucre bract, compact to open; heads 3-many, many-flowered; perianth greenish brown to tan, 4-5 mm long, segments acuminate; stamens 6, anthers shorter than filaments; capsules ovoid, about equaling perianth, usually apex rounded with a shallow notch; seeds minutely appendaged at each end. Wet soil along roads and openings in southern and east-central Alaska.

Juncus triglumis (Three-flowered Rush) Perennial; culms densely tufted, 5-15 cm tall, circular in transverse section or nearly so; leaves 1-3 per stem, basal or nearly so, blades thread-like, circular in transverse section; inflorescence terminal, commonly almost equal to lower bract; heads solitary, 1-3(5)-flowered; perianth pale brown or whitish, 4-5 mm long, segments obtuse; stamens 6, anthers shorter than filaments; capsules cylinder-like, almost equal to or longer than perianth, acute apically; seeds less than 2 mm long including the tail. Moist sites in arctic and alpine tundra in most of Alaska. Flowers June-July.

KEY TO SPECIES OF *Luzula:*

1a. Inflorescence with solitary (rarely 2-3) flowers terminating ultimate branchlets; pedicels commonly more than 5 mm long (at least some). **(2)**

1b. Inflorescence with dense spiked or headed clusters of flowers terminating ultimate branchlets; pedicels commonly less than 1 mm long. **(4)**

    2a. Inflorescence commonly simple; pedicels almost arranged in umbels, rising from near a common point but of differing lengths.   ***L. rufescens***

    2b. Inflorescence commonly few- to many-branched; pedicels almost umbellate or spiked to raceme-like. **(3)**

    3a. Stem leaves usually 2-3, rarely more than 3 mm broad; bracts and bracteoles of inflorescence fringed with long hairs.   ***L. spadicea***

    3b. Stem leaves usually more than 3, commonly 6-10 mm broad or more; bracts and bracteoles merely appearing gnawed or shredded, usually without long fringe of hairs.   ***L. parviflora***

4a. Lowermost bract of inflorescence well developed, leaf-like, callous-thickened and rounded apically; bracts and bracteoles fringed with a few long hairs or none; seeds with an appendage near scar of seed near its point of attachment.   ***L. campestris***

4b. Lowermost bract of inflorescence poorly developed, not leaf-like or if so then awl-shaped apically; bracts and bracteoles commonly fringed with long hairs; seeds not as above. **(5)**

5a. Bracts longer than flowers, silvery, conspicuous; inflorescence on a spike, commonly nodding.   ***L. spicata***

5b. Bracts equaling or shorter than flowers, conspicuous or not; inflorescence of 1-several almost headed clusters, erect or nodding.   ***L. arcuata***

Luzula campestris Many-flowered Wood-rush

Tufted from conspicuous rhizomes, 10-60 cm tall. Culms not cespitose, decumbent, 10-20 cm. Leaf blades flat or channeled apically, 1.5-6.5 mm broad, long-hairy marginally, stem leaves 2-4, tips obtuse, callous-thickened, usually chestnut brown, basal leaf sheaths brownish to straw-colored. Inflorescence 1-5, densely congested flower clusters in heads, lowermost bract leaf-like, blade well developed, callous-tipped, bracts shorter than flowers; perianth 2.2-4.7 mm long, chestnut brown to silvery white, margins thin and transparent, attenuate or acute; anthers 0.4-1 mm long or more, from shorter to longer than filaments. Capsule ovoid, shorter than or almost equal to perianth; seeds reddish to brown, globose, 1-1.3 mm long.

Most of Alaska. Wet to dry sites in tundra, woods and thickets. Common. Flowers May-August. Also known as *L. multiflora.*

Luzula parviflora Small-flowered Wood-rush

Tufted or stems solitary from long rhizomes, 15-80 cm tall. Leaf blades flat throughout or channeled apically, 8-11 mm broad, sparingly hairy or hairless marginally; stem leaves 3-5, tips awl-shaped, often reddish or purplish; basal leaf sheaths brown or less commonly chestnut or purplish. Inflorescence congested to open nodding or spreading panicle, flowers borne singly or in pairs on elongate pedicels; lowermost bract leaf-like or blade missing; bracts shorter than flowers, appearing gnawed and often somewhat ciliate; perianth 1.5-2.2 mm long, brownish, acute apically; anthers 0.4-0.6 mm long, commonly shorter than filaments. Capsule ovoid, as long as or longer than perianth; seeds 1-1.4 mm long, brownish with cottony fibers at lower ends.

Most of Alaska. Forests and meadows at most elevations. Common. Flowers June-August.

Luzula arcuata (Curved Wood-rush) Tufted, from short rhizomes, 8-20 cm tall; leaf blades rolled under, channeled or flat, 1-3 mm broad, sparingly hairy or smooth; stem leaves 1-3, tips often brown or purplish; basal leaf sheaths brown to straw-colored or purplish; inflorescence of 3-10 or more headed or spike-like flower clusters borne on very slender drooping or arching branches; lowermost bract bladeless, sheathing, chestnut to purplish brown; bracts shorter than flowers; perianth 1.8-3.5 mm long, brown, thin and transparent apically, sharp pointed; anther shorter than filaments; seeds pointed at each end attached to tuft of white fibers. Stream beds, ridge tops and slopes in tundra at high elevation in most of Alaska. Flowers June-July.

Luzula spicata (Spiked Wood-rush) Stems tufted, 1-3 dm tall; leaf blades channeled, long-hairy marginally, 1.5-3 mm broad (folded); stem leaves 1-3, tips awl-shaped, commonly purplish or chestnut; basal leaf sheaths brown to tan or chestnut, not purplish; inflorescence of 1(2-3) congested or spike-like heads, flower clusters commonly nodding, almost stalkless or if more than one, one on a peduncle; lowermost bracts bladeless or with a slender awl-shaped blade; bracts surpassing flowers, midvein often bristled, commonly with long hairs; perianth 2.5-3 mm long; anther shorter than or equal to filaments; capsule ovoid, seeds brown with slight appendages apically. Alpine tundra, along streams and in meadows from southeastern Alaska to Aleutians. Flowers in July.

Luzula spadicea (Wahlenberg Woodrush) Tufted from short rhizomes 10-40 cm long; leaf blades flat throughout or channeled apically, 1-3 rarely 6 mm broad, sparingly hairy or hairless; stem leaves 2-3, tips awl-shaped, often purplish or brownish; basal leaf sheaths brown to chestnut or straw-colored; inflorescence open, nodding or spreading panicle, flowers solitary or in twos or threes at end of thread-like branches; bracts shorter than flowers, both shredded and ciliate; perianth 1.5-2.4 mm long, segments brownish to chestnut with thin transparent apex, acute; capsule ovoid, brown, seeds brown slightly appendaged apically. Pond margins, wet slopes, stream banks and moist meadows in arctic and alpine tundra and open woods in much of Alaska. Flowers in July. Also known as *L. wahlenbergii*

JUNCAGINACEAE
ARROWGRASS FAMILY

Triglochin maritimum   Maritime Arrowgrass

Thick-scaped herb 20-120 cm tall from woody rhizome, covered with fibrous, whitish old leaf bases. Leaves fleshy, linear, flattened or channeled. Inflorescence scape often purple near base, mostly exceeding leaves; racemes several to many-flowered, often more than half total length of plant, closely flowered. Petals greenish or yellowish. Fruit oblong to ovoid capsule obtuse at base with 6 carpels.

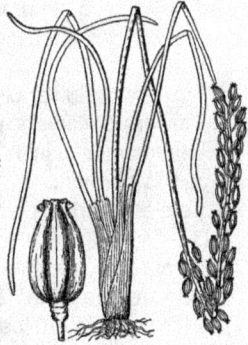

Much of Alaska except central, west-central and Aleutians. Meadows, brackish marshes, beaches and tidal flats. Common. Flowers July-August. Contains cyanide-producing glycosides. Grows with and can be confused with *Plantago maritima* (Goose-tongue) which is edible.

Triglochin palustre (Marsh Arrowgrass) Linear, less than 30 cm tall, leaves more sharply pointed than *T. maritimum*; racemes about half length of plant, perianth segments greenish or yellowish but can be tinged with purple; fruit tapering to a point, with 3 carpels. Bogs, tide flats and lake shores in most of Alaska south of Brooks Range except western portion. Also called *T. palustris*.

LILIACEAE
LILY FAMILY

KEY TO GENERA OF *LILIACEAE:*

1a. Inflorescence umbellate; plants with odor of onion.　　　　*Allium*

1b. Inflorescence racemose, paniculate, cymose, scapose, or flowers axillary; plants not onion scented. **(2)**

2a. Plants (0.5)1-2.5 m tall; stem leaves mostly 5-30 cm long, bases sheathing; flowers in large panicles (these usually over 20 cm long) or rarely reduced to a spike-like raceme.　　　　*Veratrum*

2b. Plants usually much less than 1 m tall; stem leaves usually less than 15 cm long or absent, usually not sheathing. **(3)**

3a. Leaves all basal or essentially so; scapes with solitary white flowers (rarely 2-flowered), shorter than leaves, or flowers several to many and blue to violet. **(4)**

3b. Leaves all on a stem, at least some (much smaller upward in *Lloydia, Tofieldia,* and *Zigadenus*); inflorescence various, but if a 1-flowered scape, leaves linear. **(5)**

　　4a. Flowers solitary, white; leaves mostly 3-6 cm broad.　*Clintonia*

　　4b. Flowers several to many, blue to violet; leaves commonly 0.5-1.5 cm broad; plants collected once along Haines Highway in southeastern Alaska where probably introduced.
　　　　　　Camassia quamash

5a. Flowers or flower clusters borne laterally in leaf axils.　　*Streptopus*

5b. Flowers or flower clusters borne in terminal inflorescences. **(6)**

6a. Leaves mostly in 1-3 whorls, mostly above middle of stem; flowers dark purplish brown.　　　　*Fritillaria*

6b. Leaves alternate or basal, not whorled; flowers not dark purplish brown. **(7)**

7a. Plants with leafy stems (basal leaves if present not conspicuously larger than stem leaves). **(8)**

7b. Plants with stem leaves much reduced upwards, basal leaves usually much larger. **(9)**

　　8a. Leaves heart-shaped, petiolate; perianth segments 4.
　　　　　　Maianthemum

　　8b. Leaves lanceolate to elliptic, almost sessile or clasping; perianth segments 6.　　　　*Smilacina*

9a. Perianth segments 2.5-5 mm long; inflorescence a spike-like raceme, plants from short rhizomes, leaves 2-ranked.　　　　*Tofieldia*

9b. Perianth segments mostly 6-12 mm long; inflorescence racemose or paniculate; plants bulbous; leaves not 2-ranked. **(10)**

10a. Inflorescence 1-2-flowered; perianth white with purplish veins, glands at base not distinctive; styles 1.　　　　*Lloydia*

10b. Inflorescence several- to many-flowered; perianth cream to yellowish, segments with greenish glands at base; styles 3.　　　　*Zigadenus*

Allium schoenoprasum   Wild Chives

Herb 15-75 cm tall, bulbs one to several, oblong-ovate within papery coating. Leaves coarse, semicircular in cross section, hollow at base. Flowers in dense nearly circular umbel, pink or rose-violet with darker veins. Capsules about half as long as perianth, seeds black, 1-2 in each of three locules.

Yakutat west to Alaska Peninsula north to Brooks Range. Meadows, fresh water shores and beaches. Common. Flowers May-July. Leaves eaten in early spring, bulbs in summer and fall. **Warning:** Do not confuse this plant with *Zygadenus elegans* which is poisonous. *Allium* has a strong onion smell while *Zygadenus* is odorless.

Clintonia uniflora   Beadlily

Perennial herb from creeping rhizomes, 10-15 cm tall. Leaves 2-4 in basal rosette, oblanceolate with long, soft, wavy hairs below, acute at both ends. Flowers solitary, rarely two, borne on scapes shorter than leaves, lobes spreading, petals white. Fruit a metallic blue berry.

Southeast Alaska. Forests and adjacent meadows. Occasional. Flowers late May-July. Fruit not considered edible.

Fritillaria camschatcensis   Chocolate Lily

Large perennial herb 20-60 cm tall from bulbs composed of several large, fleshy scales with numerous rice-like bulbets. Leaves mostly in 2-3 whorls of 5-9 blunt, linear leaves with a few leaves scattered near stem tip. Flowers large, nodding bells with a strong obnoxious odor. Petals dark brown, tinged green to greenish yellow on outside. Capsule six-angled, wingless, seeds flat.

Southeastern Alaska west to Aleutians, rarely some distance inland from southern coast. Meadows, bogs and tide flats, low to subalpine elevations. Common. Flowers May-July. Bulbs eaten raw, boiled, dried, or pounded into flour; a major food for indigenous peoples of Alaska. Fetid odor of flowers attract flies as pollinators.

Lloydia serotina   **Alp Lily**

Perennial herb 4-15 cm tall from small bulbs covered with gray fibrous coat, slender stems. Basal leaves linear, stem leaves alternate, reduced upwards. Flowers 1-few terminal, stamens 6, pistil 1 with a single style, stigma 3-lobed. Petals creamy white with purplish midveins and nerves tinged with rose on back. Fruit 3-loculed capsule, rounded, bluntly triangular in outline.

Most of Alaska. Rocky places, subalpine and alpine meadows. Occasional. Flowers June-July.

Maianthemum dilatatum   **False Lily-of-the-valley**

Perennial herb 10-40 cm tall, usually in large colonies, breaks 3-6 rule of lilies by having 4 petals. Stem leaves two to three, shiny, thick, evergreen, heart-shaped, parallel-veined. Raceme with many tiny flowers rising above leaves. Flower color cream to white. Berries spotted, becoming red when dry.

Southeastern Alaska west to Aleutians. Forests, forest edge, open grassy beach forest, bogs, along streams and lakeshores, low to mid elevations. Often the dominant groundcover species in Sitka Spruce forests. Common. Flowers late May to early July. Leaves rubbed on sore eyes. Berries probably not edible; plant contains glycosides which are heart stimulants, poultice made of whole or mashed leaves applied to boils, burns, cuts and wounds. The fruit used as medicine in the treatment of tuberculosis. An infusion of the pounded roots has been used as a wash for sore eyes, chewed roots used as a poultice on the eyes.

Smilacina racemosa   **False Solomon-seal**

Perennial herb 30-100 cm tall, stems from fleshy rhizome, slightly zigzagged. Leaves parallel veined on arching stem, oblong-lance-shaped, stalkless or short stalked with short stiff hairs below, margins minutely hairy. Flowers numerous, star-like, flowers in branched raceme-like inflorescence, white or greenish white. Fruit mottled berry becoming red at maturity.

Extreme southeastern Alaska. Forests. More common south of our range. Flowers May-July. Berries edible but not all that tasty. Roots boiled as a tea and used for rheumatism, back and kidney troubles. New name is *Maianthemum racemosum*.

Smilacina stellata   Starry Solomon-seal

Perennial herb 30-50 cm tall, rhizome forking. Leaves oblanceolate to elliptic, smooth, stalkless, hairy beneath. Flowers numerous in star-like terminal clusters, unbranched, white or greenish white. Berries red-purple, becoming black.

Extreme northeast southeastern Alaska and south-central Alaska. Meadows, open forests and bogs. Uncommon. Flowers June. Berry edible but not palatable. New name is **Maianthemum stellatum**.

Streptopus amplexifolius   Wild Cucumber

Perennial herb 30-100 cm tall, erect, stem usually branched, stem base usually hairy. Leaves stalkless, clasping stem, ovate to lance-shaped, alternate, prominent veins. Petals bell-shaped, hanging from slender, kinked stems below leaves, yellowish white, rarely pinkish. Fruit yellow when immature, red to black when mature; oblong berry.

Most of Alaska south of 65th parallel. Forests and forest edge. Common. Flowers May-early July. Young succulent shoots taste like fresh cucumber; berries edible but act as a laxative. Hitchcock(1974) recognizes two distinct subspecies: **S. a. americanus** coarsely hairy stems at base; and a glabrous subspecies (**S. a. chalazatus**). Hulten (1968) recognizes in addition two other subspecies with intermediate degrees of hairiness. In our area **S. a. americanus** generally dominant.

Streptopus roseus   Rosy Twisted-stalk

Perennial herb 10-50 cm tall, stem simple and zigzagged. Leaves ovate to lance-shaped, rounded to slightly clasping at base, veins often prominent; margins finely hairy. Petals and sepals bell-shaped from peduncles below upper leaves, rose-colored to pinkish, petal tips fringed white. Fruit mature berries red or yellow, often deltoid in shape, semi-translucent.

Southeastern Alaska west to Prince William Sound. Open forests, small streams and forest edges to subalpine. Common. Flowers late May-July. Berries edible but if consumed in quantity, diarrhea symptoms can result. Also known as **S. lanceolatus.**

225

Streptopus streptopoides Small Twisted-stalk

Herb 3-20 cm tall, stem simple with upper stem often prostrate. Leaves unstalked, smooth, ovate-lance-shaped, unclasped to somewhat clasping at base. Flowers saucer-shaped, borne singly in axils of upper leaves, petals united basally forming a short tube, spreading, petals purplish-green. Fruit a red globular berry.

Southeastern to south-central Alaska. Dense conifer forests. Common. Flowers May-early June. Also known as *Kruhsea streptopoides.*

Tofieldia glutinosa   Sticky Tofieldia

Perennial herb 15-40 cm tall with 2-3 linear leaves near base. Leafless flower stem covered with minute, reddish, glandular hairs, flowers in terminal spike-like racemes, white or yellowish-green. Fruit 3-loculed capsule 4-7 mm long, usually tinged reddish-purple.

Southeastern Alaska west to Kenai Peninsula. Meadows, bogs and along shores in Prince William Sound. Common. Flowers June-July.

Tofieldia pusilla (Scotch Asphodel) Tufted, stem with at least 1 poorly developed, usually nearly basal stem leaf, leaves folded together lengthwise in two ranks; flowers yellowish white to greenish; fruit green. Arctic and alpine tundra to open woods from Juneau mainland area north through most of continental Alaska. Flowers June-July. *T. coccinea* has one or two leaves on scape; flowers whitish tinged with red; range similar to that of *T. pusilla.*

Veratrum viride   False Hellebore

Stout, leafy perennial 70-200 cm tall. Leaves broadly round-oval to ovate-lance-shaped, strongly parallel-veined, base of leaf clasping stem, leaves woolly below. Flowers star-shaped, petals pale green with dark-green centers on long, drooping stems. Fruit 3-loculed oval or oblong capsule splitting down septa.

SE Alaska west to Aleutians north to Nome. Meadows, moist places and subalpine meadows. Common. Flowers May-August. **Warning:** Contains poisonous alkaloid, causes vomiting, purging, paralysis and death from asphyxia. Used by Tlingit people as a strong but respected medicine for many illnesses. Also known as *V. eschscholtzii.*

Zygadenus elegans   Elegant Death Camas

Perennial herb 20-70 cm tall, bulbous, outer coat of bulb fibrous and smooth. Leaves mainly basal, stem with 1-2 grass-like leaves, keeled, covered with a whitish waxy coating. Flowers in loose cylindrical raceme, middle and upper bracts translucent-margined, greenish or yellowish white. Fruit 3-loculed capsule that splits down each partition, twice as long as perianth. Seeds numerous, straw-colored, linear-ellipsoid, about 5 mm long.

Northeast southeastern Alaska throughout most of Alaska except Aleutians. Open forest, alpine meadows, dry areas and along trails. Common farther north, but uncommon in Southeast Alaska. Flowers late June-August.

Warning: All parts of this plant are poisonous if ingested. It contains the alkaloid zygadenine which causes vomiting, lowered body temperature, difficult breathing and coma.

ORCHIDACEAE
ORCHID FAMILY

KEY TO GENERA OF *ORCHIDACEAE:*

1a. Leaves absent at flowering time; plants reddish or yellowish saprophytes with coral-like roots.   *Corallorhiza*

1b. Leaves present at flowering time; plants greenish with roots various but not coral-like. **(2)**

2a. Fertile stamens 2; lip bag-shaped, 1.2-3 cm long; leaves 2-several.   *Cypripedium*

2b. Fertile stamens 1; lip usually neither bag-shaped nor over 1.5 cm long (except in *Calypso*, which has a single leaf). **(3)**

3a. Lip of corolla (not of calyx) bearing a prominent spur at base.   *Platanthera*

3b. Lip of corolla lacking a spur at base. **(4)**

4a. Leaves 1, folded lengthwise into prominently ribbed plaits; lip bag-shaped, 1.5-2.5 cm long.   *Calypso*

4b. Leaves (1)2-several, variously folded or flat but not plaited; lip usually not bag-shaped, less than 1.5 cm long. **(5)**

5a. Plants with small bulbous corms; pollen grains smooth or waxy.   *Malaxis*

5b. Plants from rhizomes or fascicled roots; pollen grains surface covered with granules or powdery. **(6)**

6a. Leaves 2, opposite; flowers mostly less than 10 in a lax raceme.   *Listera*

6b. Leaves several, alternate or basal; flowers mostly more than 10 in a compact or lax raceme or spike. **(7)**

7a. Roots from a creeping rhizome; basal leaves broadly elliptic to lance-shaped or ovate, greenish above, pale beneath.   *Goodyera*

7b. Roots from base of stem; basal leaves mostly narrowly lance-shaped, not markedly different above and below.   *Spiranthes*

Calypso bulbosa Calypso Orchid

Erect, hairless perennial 10-25 cm tall from fleshy subterranean bulb-like stem with one soon-wilting basal leaf and 2-3 brown sheaths. Leaves long-stemmed, blades heart-shaped-ovate to elliptic, folded in fanlike pleats. Flowers solitary, large with mostly pink, sharp pointed perianth leaves, broad, pale, purple-spotted lip, several color forms occur, petals variegated with purple, pink, and yellow. Fruit erect, ellipsoid capsule.

Southeastern and disjunctly in interior Alaska. Deeply shaded and beach fringe forests especially on small islands. Uncommon to rare in southeastern Alaska. Flowers late May-June. Forms mycorrhizal associations. Digging roots (corms) can extinct our localized populations.

Corallorhiza mertensiana Western Coralroot

Perennial saprophyte 15-60 cm tall from a coral-like rhizome, plant reddish or brownish-purple, erect, hairless, with 2-3 sheathing bracts on flowering stalk. Green leaves reduced to thin, semi-transparent sheaths. Flowers 1-25 purplish, on solitary stalk, sepals 3-nerved, widely spreading, lip of petal mottled reddish purple, oblong to broadly obovate, 3-nerved with a pair of teeth near base, prominent rounded projection on underside of ovary immediately behind purplish petals. Fruit capsule 15-18 mm long.

Southeastern Alaska to Prince William Sound. Coniferous forests. Occasional. Flowers July-August. Also known as *C. maculata*.

Corallorhiza trifida Early Coralroot

Saprophytic herbs 10-30 cm tall, stem erect, hairless and leafless, pale yellow, greenish or brownish tinged, with 2-3 sheathing bracts, bract apex often flaring and brownish. Flowers 3-20 yellowish white to purplish, brownish, or greenish, sepals 1-nerved, lip white, round to obtuse at apex, spotted with purple. Capsule 8-12 mm long.

Northern southeastern Alaska west to Aleutians north to Fairbanks. Forests, sometimes along streams or seeps. Occasional to rare in southeastern Alaska, more common in interior Alaska. Flowers June-July. Paler and smaller than *C. mertensiana*.

Cypripedium calceolus   Yellow Ladyslipper

Perennial 10-30 cm tall from slender rhizome, stems glandular-hairy with yellowish crosswalls and 3-5 sheathing leaves. Stem leaves 3-5, hairy on veins and margins. Flowers 1-3 in axis of leaf-like bract, fragrant sepals usually purplish, upper ovate, lower united, longer or almost equal to lip, petals twisted spirally, lip yellow, purple-veined, pouch-shaped. Capsule ellipsoid, glandular-hairy.

Northern southeastern Alaska and central western Alaska. Forests. Rare. Flowers May-June.

Cypripedium montanum (Mountain Ladyslipper) Perennial 20-70 cm tall from a stout rhizome, stems pubescent with multicellular hairs with yellowish crosswalls, 4-6 alternate ovate leaves; 1-3 flowers, petals brownish purple or dark green tinged with purple, lip mostly white, tinged or veined with purple. Open woods at high elevations in southeastern Alaska in Glacier Bay, near Haines Highway and Hope Highway in south-central Alaska. Flowers May-early July. It takes 15 years for this orchid to flower.

Goodyera oblongifolia   Rattlesnake Plantain

Evergreen perennial 20-45 cm tall with stout stem, glandular above, from short rootstock with fibrous roots. Leaves in basal rosette, oblong to elliptic, dark green or marked with white especially along midvein. Raceme one-sided, flowers several to many, perianth segments glandular, petals and dorsal sepals forming hood, lip pouch-like, petals white, tinged green. Fruit an erect capsule.

Southeastern to south-central Alaska. Beach fringe, dense forests and primary glacial forests, low to mid elevations. Occasional. Flowers July-August. Leaves cut open and moist inner parts used on wounds.

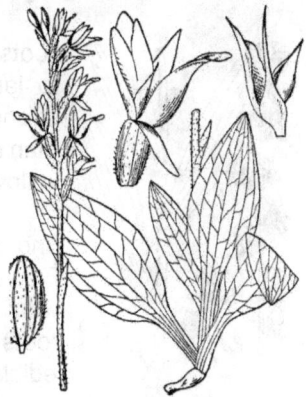

KEY TO SPECIES OF *PLATANTHERA* AND *PIPERIA:*

1a. Lip unequally three-toothed at apex, middle tooth small.　　*P. viridis*
1b. Lip entire at apex. **(2)**
2a. Leaves 1-2 (rarely 3), basal or essentially so; stems without bracts or with a single one (rarely more) near middle. **(3)**
2b. Leaves several, on stems or almost basal; stems leafy or with conspicuous bracts. **(5)**
　　3a. Perianth segments short, 1.5-2.5 mm long; spur bulbous.
　　　　　　　　　　　　　　　　　　　　　P. chorisiana
　　3b. Perianth segments mostly over 2.5 mm long; spur elongate. **(4)**
　　4a. Leaves 2 (rarely 3), large, orbicular to oval, prostrate; spur over 14 mm long.　　*P. orbiculata*
　　4b. Leaves 1-2, small, obovate to linear-oblanceolate, erect or spreading; spur less than 12 mm long.　　*P. obtusata*

5a. Leaves clustered at or near base of stem, usually withered at flowering; stems with numerous bracts; lip ending abruptly at base or angled at each side; sepals 1-nerved.　　***Piperia unalaschensis***
5b. Leaves on stem, persisting at flowering time; stem with leaf-like bracts; lip not ending abruptly at base; sepals 3-nerved. **(6)**
6a. Lip ovate-lance-shaped, abruptly dilated at base; flowers white, rarely greenish.　　*P. dilatata*
6b. Lip linear to broadly lance-shaped, not dilated at base; flowers usually greenish (sometimes purplish tinged). **(7)**
7a. Spur bulbous to strongly pouch-shaped; lip linear to lance-shaped; raceme usually laxly flowered, elongated.　　*P. saccata*
7b. Spur cylindrical, sometimes club-shaped; lip elliptical to lance-shaped; raceme usually short to congested.　　*P. hyperborea*

Platanthera dilatata　　White Bog-orchid

Perennial 15-70 cm or taller, thick, fleshy roots, stem leafy. Leaves ovate-lance-shaped to lance-shaped. Flowers sweet scented, lip strongly dilated at base, spur cylindrical, shorter than or nearly equal to length of lip. Petals white, yellowish, or greenish. Fruit an elliptic capsule.

Southeastern Alaska west to Aleutians. Bogs and wet meadows. Most common *Platanthera*. Flowers late June-August. A narrow-leaf form with spur longer than lip common. Tuber-like roots eaten raw or cooked, but its status as a safe edible needs clarification. *Platanthera* sometimes placed in genus *Habenaria*.

Platanthera saccata   Slender Bog-orchid

Herb 15-100 cm tall or more from thick fleshy roots, stems leafy. Leaves narrowly lance-shaped to elliptic. Raceme with few to many flowers, flower lip linear to oblong, not expanded at base, often dark colored, longer than spur, spur pouch-like. Petals green, tinged purple brown. Fruit a capsule.

Southeastern Alaska west to eastern Aleutians. Bogs and meadows low to mid elevations. Common. Flowers July-August. Also known as *P. stricta*; includes *Plantanthera gracilis*. *P. gracilis* is similar to *P. saccata* but with a thread-like spur longer than lip; found in wet meadows in southern Southeast Alaska.

Platanthera viridis   Frog Bog-orchid

Perennial from thick fleshy finger-like lobed tuberoids, stem flimsy. Lower stem leaves close to stem base. Flowers few, inflated spur and lip 2-3-lobed, sometimes entire, remaining 5 perianth leaves form a hood over lip. Petals green or brown. Fruit a capsule.

Southeastern Alaska west to Aleutians and western Alaska. Alpine meadows. Occasional. Flowers middle June-August. Also known as *Coeloglossum viride*.

Platanthera chorisiana (Chorus Bog-orchid) 7-25 cm tall with 2 (rarely 1) almost basal, ovate to elliptic leaves, solitary (rarely 2) stem bracts; racemes few to several flowered, flowers yellowish, lip entire, spur bulbous. Dry ridges and bogs from southeastern Alaska to Aleutians, uncommon. Flowers July-August.

Platanthera hyperborea (Northern Bog-orchid) 15-100 cm tall; racemes with few to many flowers, flowers green or yellowish, lip linear to lance-shaped, not dilated, spur usually shorter than lip. Stream beds, lake margins, seeps, meadows and open woods in southern and west-central Alaska. Flowers July-August. Includes *Platanthera convallariifolia*.

231

Platanthera obtusata (Small Bog-orchid) 6-35 cm tall, stems leafless or with bracts, basal leaves solitary or rarely two, obovate to oblanceolate; flowers greenish white, lip linear-lance-shaped, spur slender, almost equal to lip. Woods, thickets and along streams in most of Alaska south of Brooks Range except Aleutians. Flowers June-July.

Platanthera orbiculata (Round-leaved Bog-orchid) 6-60 cm tall, 2 round leaves and greenish white flowers, sepals strongly nerved, lip linear-oblong, spur longer than lip. Woods in extreme southeastern Alaska. Flowers June-July.

Platanthera unalaschensis (Alaska Bog-orchid) 25-90 cm tall, stems tan to purplish brown, 2-4 oblanceolate to narrowly lance-shaped, leaves near or at base, usually withered by flowering time; flowers white to yellowish green, often marked with purple, lip ovate to lance-shaped, spur cylindrical, about equal to lip. Flowers July-August. Bogs and meadows from southeastern Alaska west to Unalaska. Sometimes included in the genus ***Habenaria*** or ***Piperia***.

Listera caurina   Northwestern Twayblade

Perennial 10-30 cm tall from creeping rhizomes. Leaves broadly oval in one opposite pair near mid-length of stem. Flowers often hairy, sepals and petals 1-nerved, lower lip petal broadly wedge-shaped with a rounded or slightly pointed tip, greenish to yellowish. Fruit an egg-shaped capsule to 6 mm long.

Southeastern Alaska to Prince William Sound. Forests. In southeastern Alaska, not as common as ***L. cordata***. Flowers June-early August.

Listera cordata   Heartleaf Twayblade

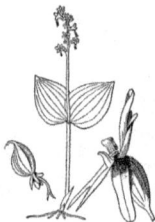

Perennial 10-20 cm tall from slender creeping rhizome. Leaf stem with single pair of broad, opposite, somewhat heart-shaped leaves. Flowers above leaves forming a long, loosely flowered spike with petals deeply divided into two linear lobes, green or purple. Fruit an ovoid capsule 4-6 mm long.

Southeastern Alaska west to Aleutian. Mossy coniferous forests. Common. Flowers June-early August.

Listera convallarioides (Broadleaved Twayblade) Somewhat stoloniferous, 5–37 cm. Stems green, succulent, glabrous. Leaves broadly ovate to elliptic, oval or suborbiculate. Inflorescences 5–20-flowered, lax. Lip of yellowish-green flowers longer, more notched with hairier margins than the other species. Along shores and streams; Stikine River area and disjointly to the Aleutians. Rare.

Malaxis monophyllos   White Adder's-tongue

Slender, inconspicuous herb to 30 cm tall arising from a subterranean bulb-like stem. Leaves 1 or 2 with a long sheath on lower part of stem. Flower in slender raceme, sepals and petals 1-nerved, lip 3-lobed, central lobe tapering to a point. Petals greenish to yellowish-green. Fruit an ovoid capsule 3-6 mm long.

Southeastern Alaska west to Aleutians. Open forest, meadows, bogs, and coast. Occasional to somewhat common. Flowers late June-July. Also known as *M. monophylla*.

Malaxis paludosa (Bog Adder's-tongue) Perennial to 20 cm tall, 2-5 leaves in a cluster with small yellowish-green flowers; egg-shaped capsules. Bogs from southeastern Alaska to south-central Alaska. Also known as *Hammarbya paludosa*.

Spiranthes romanzoffiana Hooded Ladies-tresses

Perennial from fleshy tuberous roots with erect stem to 50 cm, stem leafy in lower part. Leaves several, mostly basal, linear to oblanceolate. Flowers fragrant, densely flowered spike, composed of three spiral ranks, sepals and two petals forming hood, lip constricted above middle, recurved near apex, bracts longer than flowers. Petals white to cream. Fruit dry many-seeded capsule.

Much of Alaska south of Brooks Range except coastal western regions. Bogs, meadows and open forests, low to mid elevations. Occasional. Flowers July-August.

233

POACEAE
GRASS FAMILY

KEYS TO GENERA OF *POACEAE:*

1a. Inflorescence a spike; spikelets sessile. **KEY 1**
1b. Inflorescence a panicle (this sometimes dense and spike-like); spikelets borne on a pedicel. **(2)**

2a. Spikelets awned; awns arising from glumes or lemmas or both.   **KEY II**
2b. Spikelets awnless. **(3)**

3a. Spikelets with a single floret. **KEY III**
3b. Spikelets with 2-several florets. **KEY IV**

KEY I. INFLORESCENCE A SPIKE; SPIKELETS SESSILE

1a. Spikelets averaging 2-3 per node of rachis. **(2)**
1b. Spikelets usually 1 at each node of rachis (sometimes with 2 per node at base).
Agropyron
2a. Spikelets 1-flowered, 3 per node, 2 lateral ones on short peduncles reduced to awns (fertile in *H. vulgare*). *Hordeum*
2b. Spikelets 2-6-flowered, usually 2-3 per joint, all alike. *Elymus*

KEY II. SPIKELETS AWNED, AWNS ARISING FROM GLUMES OR LEMMAS OR BOTH

1a. Spikelets with a single floret. **(2)**
1b. Spikelets with 2-several florets. **(5)**

2a. Panicle very dense, spike-like, cylindrical, ovoid, or globe-like. **(3)**
2b. Panicle open or contracted, but not forming dense heads. **(4)**

3a. Glumes awned, stiffly hairy on keels; lemmas awnless.
Phleum
3b. Glumes bent, awnless, softly hairy on keels; lemmas awned.
Alopecurus
4a. Florets with conspicuous, long hairs at base; palea present, rachilla prolonged behind it. *Calamagrostis*
4b. Floret naked at base, or with short hairs only; palea lacking in some species. *Agrostis*
5a. Spikelets with 2 hairy rudimentary lemmas below single perfect flower; lemmas hard and shiny, pale yellowish. *Hierochloe*
5b. Spikelets without rudimentary lemmas below; imperfect florets, when present, borne below perfect ones. **(6)**

6a. Glumes shorter than first floret; lemmas awned from tip or from bifid apex. **(7)**
6b. Glumes (at least one) as long as lowest spikelet or as long as entire spikelet; lemmas variously awned. **(8)**

    7a. Lemmas awned from tip.   *Festuca*
    7b. Lemmas awned from below tip, between teeth of a bifid apex. *Bromus*

8a. Ligules a fringe of hairs; lemmas bifid at apex, awned from between lobes.
    Danthonia
8b. Ligules various, but not a fringe of hairs; lemmas awned from back, toothed apically. (9)
9a. Lemmas keeled, awned from above middle.   *Trisetum*
9b. Lemmas convex, awned from above or below middle.   *Deschampsia*

KEY III. SPIKELETS AWNLESS, WITH A SINGLE FLORET.
1a. Spikelets with 2 slender hairy rudimentary lemmas below single perfect floret; lemmas hard and shiny, pale yellowish.   *Phalaris*
1b. Spikelets without rudimentary lemmas below; imperfect florets, if any, above perfect ones. (2)

2a. Florets on short but distinct pedicels; rachilla prolonged behind palea; tall drooping plants of damp or wet shady places.   *Cinna*
2b. Florets sessile; rachilla not prolonged behind palea (except in some *Agrostis*). (3)

3a. Plants usually 4-2 cm tall, rarely over 15 d cm; glumes usually much shorter than lemma.   *Phippsia*
3b. Plants 20-60 cm tall or more; glumes longer or shorter than lemma. (4)

4a. Glumes longer than lemma; spikelets mostly less than 3 mm long.   *Agrostis*
4b. Glumes shorter than lemma; spikelets over 3 mm long.   *Arctagrostis*

KEY IV. SPIKELETS AWNLESS, WITH 2-SEVERAL FLORETS
1a. Glumes equaling or longer than spikelet; slender plants of wet soil.   *Hierochloe*
1b. Glumes shorter than spikelet; plants of various habitats. (2)

2a. Glumes papery; lemmas firm, strongly nerved, rough-margined; spikelets tawny or purplish, usually not green.   *Melica*
2b. Glumes nor papery; lemmas and spikelets various, but not as above, or seldom so. (3)

3a. Nerves of lemma converging at summit. (4)
3b. Nerves of lemma parallel or nearly so, not converging at summit. (6)

    4a. Lemmas keeled or rarely rounded on back, awnless; leaf tips boat shaped.   *Poa*
    4b. Lemmas rounded on back, often shortly awned; leaf tips not or only slightly boat-shaped. (5)

    5a. Lemmas acute or more or less obtuse or awned, awn inserted near apex; leaf sheaths open.   *Festuca*
    5b. Lemmas acute to obtuse, usually awned from below tip (usually between teeth of a bifid apex); leaf sheaths closed (usually to near top).
    Bromus
6a. Nerves of lemmas prominent; glumes much shorter than adjacent lemma.
    Glyceria
6b. Nerves of lemmas inconspicuous; glumes much shorter or nearly as long as adjacent lemma.   *Puccinellia*

Agropyron trachycaulum   Bearded Wheatgrass

Tufted grass, rhizomes lacking or sometimes short, 25-90 cm tall, nodes of culms hairless or less commonly hairy. Ligules entire or irregularly toothed to fringed with hairs, to 0.8 mm long, blades usually flat or less commonly more or less rolled, mostly 1.5-6 mm broad, rough to touch, hairless or less commonly hairy above. Spikes 4-21 cm long, more or less compact; spikelets usually much longer than rachis internodes, 9-18(22) mm long (excluding awns), 3-6-flowered, rachilla rarely exposed; glumes three-fourths to about as long as spikelet, lance-shaped to oblanceolate, thin, transparent margins, 3-7-ribbed, acute to slender awned; lemmas hairless or slightly rough to touch, or less commonly minutely hairy; awnless to awned (1-3)4-20(25) mm long, straight or curved; anthers 1-2 mm long.

Most of Alaska except Aleutians. Meadows, woodlands, gravel bars, beaches and talus slopes. Occasional. Flowers June-July. Occurs in a vast number of forms and apparently susceptible to modifications by soil and water conditions. Includes *A. boreale, A. macrourum* and *A. caninum.* Known to hybridize with *Hordeum jabatum* resulting in *Agrohordeum macounii*.

KEY TO SPECIES OF *AGROSTIS:*
1a. Palea well developed, usually at least half as long as lemma. **(2)**
1b. Palea lacking, or if present never as much as half length of lemma. **(5)**
     2a. Anthers 0.5-0.6 mm long; plants usually less than 30 cm tall.
                                                                                A. thurberiana
     2b. Anthers mostly 1-1.4(1.8) mm long; plants often over 30 cm tall. **(3)**
     3a. Rachilla prolonged behind palea as a short stub or bristle 0.5-1 mm long; plants uncommon, on coasts and islands of southern Alaska.
                                                                                A. aequivalvis
     3b. Rachilla obsolete, not evident as a stub or bristle; plants with various distribution. (4)
     4a. Lemmas with exserted abruptly bent awns.   *A. mertensii*
     4b. Lemmas awnless or with included or only slightly exserted, usually straight awns. **(5)**
5a. Panicle dense, interrupted, at least some of lower branches spikelet-bearing from base; plants of coasts and islands of southern Alaska.   *A. exarata*
5b. Panicle loose, open, sometimes diffuse, none of branches spikelet bearing from base. **(6)**
6a. Anthers 0.5-0.8 mm long or more; awns lacking or very small; plants of coasts and islands of southern Alaska, seldom in Interior.   *A. alaskana*
6b. Anthers less than 0.5(0.6) mm long; awns lacking or present; plants widespread.
                                                                                A. scabra

Agrostis exarata   Spike Redtop

Perennial grass, rhizomatous or rhizomes not developed, sometimes with stolons, culm to 100 cm tall, tufted. Ligules 2-10 mm long or more; leaf blades flat, (1)2-10 mm broad, rough to touch. Panicle contracted, commonly interrupted, 5-25 cm long or more, branches ascending to erect, green to purplish; glumes unequal, first commonly 2-3.5 mm long; lemmas shorter than glumes, awnless or rarely awned; palea about one-third as long as lemma, callus slightly hairy; rachilla joint obsolete; anthers 0.4-0.7 mm long.

Southeastern Alaska west to Aleutians. Open areas at low elevation, ocean beaches, meadows and open woods. Common. Flowers June-August.

Agrostis scabra   Ticklegrass

Perennial grass to 80 cm tall, tufted with numerous narrow leaves at base. Ligules 2-4 mm long, rarely longer; leaf blades flat or folded, 0.5-2(3) mm broad, slightly rough to touch. Panicle at first narrow, in mature state diffuse, 6-30 cm long, branches spreading to erect, purplish; glumes unequal, first 2.1-3.0 mm long; lemmas shorter than glumes, awnless or awned, awn straight or nearly so; anthers 0.3-0.6 mm long.

Throughout Alaska. Open areas. Common. Flowers July-August.

Agrostis aequivalvis (Northern Bentgrass) 30-80 cm tall, forming small tufts; ligules 1-5 mm long, blades nearly smooth, inflorescence a narrow panicle, often purple, branches in half whorls; spikelets 3-4.5 mm long; glumes about 3 mm long, hairless on keel, slightly roughened near tip, palea nearly as long as lemma. Bogs and lake margins uncommonly on coasts and islands from southeastern Alaska to Aleutians. Also known as *Podagrostis aequivalvis*.

Agrostis alaskana (Alaska Bentgrass) Perennial, shortly rhizomatous, mostly 10-80 cm tall, forming small to large tufts; ligules 2-6 mm long; leaf blades flat to rolled, minutely roughened; panicle loose with spreading or ascending branches, 2-16 cm long or more, purplish or greenish purple; glumes slightly unequal, awnless or with a straight awn shorter than spikelet or less commonly protruding; palea lacking or minute. Bogs, lake shores and ocean beaches from southeastern Alaska west to Aleutians.

Agrostis mertensii (Red Bentgrass) Perennial, tufted, 10-35 cm tall; ligules 1-3 mm long, leaf blades flat to rolled, mostly 1-3 mm broad, minutely roughened; panicles loose with spreading or ascending branches, 2-12 cm long, purplish; glumes unequal to almost equal; lemmas shorter than glumes, awned, awn curved or twisted, protruding from spikelet, callus hairy; palea minute. Stream banks, bars and open slopes mostly in arctic and alpine tundra in most of Alaska south of Brooks Range. Also known as ***A. borealis***.

Agrostis thurberiana (Thurber Bentgrass) Perennial, tufted, rhizomatous, 10-30 cm tall, ligules 1-3 mm long, leaf blades flat, 0.5-3 mm broad; panicles loose, branches ascending, purple to greenish; glumes about equal, lemma about equaling shorter glume, awnless. Wet meadows mostly in alpine from southeastern Alaska to eastern Aleutians. Also known as ***Podagrostis thurberiana***.

Alopecurus aequalis Shortawn Foxtail

Perennial grass, sometimes flowering first year, tufted, erect or decumbent and rooting at nodes, 10-60 cm tall. Ligules 3-5 mm long or more, leaf blades usually flat, 1-5 mm broad, rough to touch (at least below). Panicle 1.4-7.5 cm long, pale green; glumes 1.9-2.4 mm long with long, soft, somewhat wavy hairs on nerves, more or less silky on back; lemma almost equal to glumes, awn inserted near middle and almost equal to glumes or somewhat longer; anthers less than 1 mm long.

Much of Alaska. Bogs, river banks, roadsides, shallow ponds and along streams. Occasional. Flowers July-August.

Arctagrostis latifolia Polargrass

Perennial, rhizomatous grass 25-140 cm tall, tufted or solitary. Ligules 3.5-8.5 mm long, irregularly cleft, often tinged with red or purplish at base; leaf blades 2-15 mm broad, rough to touch. Panicles contracted to loose and open, 5-40 cm long, yellowish or green or commonly purplish; glumes unequal, first 1.7-4.7 mm long; lemma longer than glumes, awnless, 3-6 mm long; palea almost equal to lemma; rachilla joint to about 1 mm long; anthers mostly 1.5-3 mm long.

Most of Alaska. Arctic and alpine tundra and woods. Common. Flowers July-August.

Bromus sitchensis   Alaska Brome

Tufted without rhizomes, 50-150 cm tall, leaf sheaths hairless or with soft hairs, lacking auricles. Ligules 2-8 mm long, blades 5-15 mm broad, hairless or with soft hairs. Panicle 10-35 cm long, branches erect to spreading or drooping, spikelets strongly compressed, narrowly lance-elliptic to oblong, (25)35-52 mm long, mostly with 5-11 flowers; glumes unequal, first 3-5-nerved and 7.5-14 mm long; lemma lance-shaped, keeled in back, with straight or moderately bent awn to 10 mm long or more; anthers to 6 mm long.

Coast and islands of southern Alaska. River banks, beaches and mountain slopes. Common. Flowers July-September.

Bromus ciliatus (Fringed Brome) Perennial, tufted, 45-120 cm tall, leaf sheaths hairy, at least at base, auricles lacking, ligules 0.4-1 mm long, leaf blades 4-12 mm broad; panicle 8-18 cm long, distinctly nodding; spikelets slightly compressed, drooping, 14-23 mm long with 5-9 flowers; glumes unequal; lemmas lance-shaped with fringe of hairs along margin, awn 1-4 mm long, palea shorter than lemma. Open woods and thickets in central and southern Alaska. Also known as **B. richardsonii**.

Bromus pacificus (Pacific Brome) Perennial, tufted, 60-170 cm tall with backward turned hairs at nodes; leaf sheaths with soft hairs, hairs more or less turned backward, lacking auricles; ligules 1.8-3.2 mm long, blades 6-16 mm wide, with soft hairs on upper surface; panicle 13-22 cm long, open with spreading to drooping branches; spikelets somewhat compressed, 20-28 mm long, with 5-11 flowers; glumes unequal, hairy, first 1 (rarely 3)- nerved and 6-8.5 mm long, lemmas lance-elliptic, silky throughout, awns 3.5-6 mm long. Moist woods and beaches on islands of southeastern Alaska.

Calamagrostis canadensis   Bluejoint

Strongly rhizomatous grass to 2 m tall, stems smooth. Leaf sheath hairless to rough to touch, ligules 2.5-12 mm long, leaf blades rolled or more commonly flat, rough on both sides. Panicle usually open, commonly more than 2 cm broad (when pressed), 6-23(25) cm long; glumes usually purplish but sometimes greenish or straw-colored, usually rough to touch, lemma distinctly to barely shorter than glumes, awned from median one-third to apical one-third of lemma, awn straight or seldom curved, shorter to slightly longer than glumes; callus hairs at least three-fourths as long and often slightly surpassing lemma; anthers 1-1.8 mm long.

Throughout Alaska. Meadows, open areas, beaches and shrub lands. Most common and widely distributed grass in our area. Flowers June-August. Includes *C. inexpansa*.

Calamagrostis nutkaensis   Pacific Reedgrass

Culms from short rhizomes, stout, strongly tufted to 90 cm tall, stems smooth or rough below inflorescence only. Leaf sheaths smooth, ligules mostly 2-4 mm long; leaf blades usually flat and 4-8 mm broad or more, or more or less rolled, often rough on both sides. Panicle narrow, more or less loose, mostly 9-20 cm long, greenish or purplish; glumes unequal, very narrow, (4.5)5-8.3 mm long, hairless or rough on keel; lemma shorter than glume, awned from median one-third, awn straight or abruptly bent from near base and exserted from side of spikelet, seldom vestigial; anthers 1.8-3 mm long.

Along coast and islands from southeastern Alaska west to Aleutians. Wet places, exposed outer coastal areas. Common. Flowers July-September.

Calamagrostis purpurascens (Purple Reedgrass) Strongly tufted and often shortly rhizomatous, 25-100 cm tall; ligules 1-6 mm long, leaf blades flat or margins rolled under, 2-4 mm broad, rough at least on upper surface; panicle congested to loose, 5-17 cm long; glumes unequal, 5-7.8 mm long, rough over back or at least on keel; lemma distinctly shorter than glume, awned from basal one-third, awn twisted and abruptly bent, exceeding spikelet by 1.5-2 mm or more. Open slopes and ridge tops in most of Alaska.

Cinna latifolia   Woodreed

Tufted shortly rhizomatous grass 80-120 cm tall or more, stems smooth; sheaths hairless or slightly rough to touch. Ligules 3-8 mm long, blades 7-15 mm broad, rough to touch on both sides. Panicles open, branches spreading to drooping, 15-30 cm long, greenish or purplish; glumes mostly 3-4 mm long, usually rough to touch at least on keel; lemma almost equal to or slightly longer than glumes, awnless or with a straight almost terminal awn to about 1 mm long; anthers 0.5-1 mm long.

Coast and islands of southern Alaska from southeastern Alaska west to Alaska Peninsula. Stream banks, talus slopes, meadows and moist woods. Occasional to common. Flowers June-August.

Danthonia intermedia   Timber Oatgrass

Tufted grass 5-25 cm tall or more. Ligules very short, with a fringe of short hairs, blades flat or margins rolled under, commonly 1-3 mm broad with soft hairs or hairless. Panicles spike-like, mostly 3-7 cm long, often with 2-5 spikelets, purplish to straw-colored; glumes almost equal, 7.5-12 mm long, obscurely 3-5-nerved; lemma mostly 7-10 mm long, callus bearded, margins hairy, hairless on back, teeth 1-2 mm long, awn to 10 mm long; anthers 3-4 mm long.

Juneau north through continental southeastern and south-central Alaska. Meadows. Uncommon. Flowers July-August.

Danthonia spicata (Poverty Oatgrass) Tufted, 20-70 cm tall; ligules very short, with a fringe of short hairs; leaf blade usually with rolled margins, 0.5-2 mm broad, hairless or with soft hairs; panicle slender, open with 3-5 spikelets, green, glumes 9-12 mm long, almost equal, lemmas 4-5 mm long. Extreme southern portion of southeastern Alaska; to be expected elsewhere.

241

Deschampsia atropurpurea   **Mt. Hairgrass**

Perennial tufted grass to 65 cm tall. Ligules obtuse to truncate, irregularly toothed to almost entire, 1-3.5 mm long; leaf blades flat, (2)3-6 mm broad, hairy to hairless. Panicle opening with age, 3-12 cm long, branches often long and curved; spikelets green becoming purplish, occasionally 3-flowered; glumes almost equal, 4-6.3 mm long, first 1-nerved, second 3-nerved, longer than uppermost floret; lemmas obscurely 3-nerved with callus hairs about half as long, awned from above middle, awn twisted, to 2.5 mm long; anthers 0.8-1.2 mm long.

Much of southern Alaska. Meadows and open forest. Common. Flowers July-August. Also known as *Vahlodea atropurpurea*.

Deschampsia caespitosa   **Tufted Hairgrass**

Perennial grass, densely tufted culm to 120 cm tall. Ligules (1)2.2-6 mm long, acute; leaf blades rolled under, folded, or sometimes flat, 1-2(3) mm broad, rough at least on upper surface. Panicle open or more or less contracted 4-26 cm long, branches spreading or ascending to almost erect; spikelets tawny to straw-colored or purplish, occasionally 3-flowered; glumes almost equal or somewhat unequal, 3.0-4.5(5.2) mm long, first 1-nerved, second 3-nerved; lemmas obscurely 5-nerved, with callus hairs about one-fourth as long, awned from near base, awn straight or slightly curved to 4 mm long; anthers (1.2)1.4-1.9 mm long.

Throughout Alaska. Bogs, shallow ponds, river banks and beaches. Common. Flowers July-September. Includes *D. beringensis*.

Elymus mollis   **Beachgrass**

Grass to 150 cm or more, culms from long, stout, creeping rootstocks with old leaves at base. Ligules to 1 mm long; auricles present or lacking, leaf blades 3-18 mm broad, flat or rolled in from edges, rough to touch at least above. Spikes erect or nearly so, 5-25(30) cm long; spikelets mostly 2 per node, overlapping (except occasionally lowermost separate), with (3)4-6 flowers, glumes lance-shaped, broader above base, mostly 3-5-nerved, about equaling spikelet, awn tipped, hairy; lemmas mostly 10-20 mm long, hairy, acuminate to bristle-tipped, but scarcely awned; anthers 4.8-9 mm long.

Islands and coasts of Alaska. Upper beaches, forming belt along shore. Common. Flowers June-August. Used for pack straps, nets and basket handles. Stems used as overlay for baskets. Also known as *E. arenarius* and *Leymus mollis*.

Elymus glaucus (Western Ryegrass) Tufted, not rhizomatous, mostly 80-150 cm tall; ligules 0.5-1.5 mm long; auricles more or less developed; leaf blades flat, 5-12 mm broad, hairless or rough to soft-haired; spikes erect, stiff, 5-13 cm long or more; spikelets mostly 2 per node, overlapping with 3-5 flowers; glumes narrowly lance-shaped, broadest some distance above base, lemmas mostly 8-12 mm long, sharp-pointed, awn tipped or with a slender straight to curved awn mostly 1-3 cm long. Open woods and meadows along coasts and islands in southern Alaska.

Elymus hirsutus (Northern Ryegrass) Tufted, not rhizomatous, mostly 50-120 cm tall; ligules 0.5-1 mm long; auricles lacking or small; leaf blades flat, 4-17 mm broad, hairless or more commonly rough-hairy; spikes erect to nodding or drooping, mostly 6-15 cm long; spikelets mostly 2 per node, overlapping but not especially compact, with 3-5 flowers; glumes narrowly lance-shaped, broadest some distance above base, 3-5-nerved, produced apically into a conspicuous awn; lemma mostly 7-10 mm long (excluding awn), with long hairs on margins. Open woods and meadows on coast and islands from southeastern Alaska to the Aleutians.

Festuca rubra   Red Fescue

Perennial grass, loosely or densely tufted to 1 m tall, stem bases with persistent sheaths, shredding into coarse fibers. Ligules less than 0.5 mm long, leaf blades folded to rolled in from margins (rarely flat), mostly less than 1.5 mm wide, hairless or slightly hairy above. Panicles compact or less commonly open, 2.5-11 cm long, often to one side of stem; spikelets 8.5-17 mm long (including awns), usually 4-7-flowered, purple to green; glumes lance-shaped, first 2-4.6 mm long, second 4-7 mm long; lemmas awned, awns to 3 mm long; anthers 2-4 mm long.

Widespread in Alaska. Tidal flats, beaches, bogs and cliffs. Common. Flowers June-August. A complex and variable species.

Festuca ovina (Sheep Fescue) Perennial, densely tufted, 7-35 cm tall; stem bases clothed with persistent sheaths, these not or seldom shedding; ligules less than 0.5 mm long; leaf blades folded to margins rolled under, less than 1 mm broad, hairless to rough; panicle usually compact, 1.5-9 cm long, often on one side of stem; spikelets 4.8-10 mm long (including awns), usually 3-4-flowered; glumes lance-shaped; lemmas hairless to rough, awned, awn to 3 mm long. Alpine and arctic tundra, ridges, cliffs, woods and gravel bars in most of Alaska. Also known as *F. brachyphylla*.

243

Festuca subulata (Bearded Fescue) Perennial, tufted, 45-120 cm tall, stem bases not especially clothed with persistent sheaths; ligules less than 1 mm long; leaf blades flat, 3-10 cm wide, hairless or more or less roughened on both sides; panicle open, 10-30 cm long or more, more or less drooping; spikelets 7-10 mm long (excluding awns), 3-6-flowered; glumes narrowly lance-shaped, sharply pointed; lemmas minutely roughened to smooth, awned, awns to 15 mm long. Stream banks, open woods and rock outcrops on coasts and islands of southeastern Alaska.

Glyceria borealis   Northern Mannagrass

Rhizomatous grass rooting at lower nodes, 60-120 cm tall; leaf sheaths hairless, open near top. Ligules 4-10 mm long; leaf blades flat or folded, 2-6 mm broad, ventral surface with minute pimple-like bumps, not roughened. Panicle mostly 15-45 cm long, narrow, branches ascending to erect; spikelets narrowly oblong to linear, nearly circular in transverse section, 10-13 mm long or more, 6-11-flowered; first glume 1.5-2.5 mm long, second 3.3.5 mm long; lemma rather prominently 7-nerved, rough to touch on nerves.

Southern Alaska. Ponds and lake margins, bogs and wet meadows. Uncommon. Flowers June-September.

Glyceria grandis (American Mannagrass) Rhizomatous, often rooting at lower nodes, 100-150 cm tall; leaf sheath hairless, closed nearly or quite to apex; ligules 4-8 mm long; leaf blades flat, 3-10 mm broad or more, smooth on both surfaces or minutely roughened above; panicle mostly 20-35 cm long, open, branches spreading to ascending; spikelets more or less flattened, oblong in outline, 4.5-9 mm long, 4-7-flowered; lemma prominently 7-nerved, minutely roughened on nerves. Ponds, stream margins, bogs and wet meadows in southern half of continental Alaska including Skagway. Also known as **G. maxima**.

Glyceria leptostachya (Davy Mannagrass) Rhizomatous, mostly 70-120 cm tall or more; leaf sheath with backward facing short rough hairs, closed except near top; ligules 6-10 mm long; leaf blades flat, 3-9 mm broad, rough to touch on both sides; panicle 20-60 cm long, loose, branches ascending to appressed; spikelets narrowly oblong to linear, nearly circular in transverse section, 11-18 mm long, 8-15-flowered; lemmas prominently 7-nerved, roughened on nerves and in internerve areas. Found rarely, in pond and lake margins in southeastern Alaska.

Hierochloe alpina   Alpine Holygrass

Tufted, shortly rhizomatous grass 15-60 cm tall; basal sheath purplish. Ligules 0.6-1 mm long, with a conspicuous terminal fringe of hairs; leaf blades with margins rolled under or upper ones greatly shortened, flat or nearly so and 1-3 mm broad. Panicle more or less contracted, (1.5)2-5(5.5) cm long, branches ascending to appressed-ascending; spikelets tawny or green with purplish margins, 5.3-7 mm long (excluding awn); glumes about as long as spikelet; lemmas of staminate florets pubescent, awned, awn of first very short, that of second twisted and abruptly bent, inserted near middle of lemma, exserted from spikelets for 2-4 mm or more; fertile lemma pubescent toward apex, unawned.

Throughout Alaska. Arctic and alpine tundra and open woods. Common. Flowers June-August. Also known as ***Anthoxanthum monticola.***

Hierochloe odorata   Vanilla Grass

Solitary or tufted, rhizomatous grass 25-70 cm tall; basal sheaths pale purplish to straw-colored. Ligules 2.5-6 mm long, inconspicuously ciliate; leaf blades flat to folded, 2-5 mm broad. Panicle loose to open, 3-13.5 cm long, branches ascending to spreading; spikelets tawny to straw-colored, 3.8-6 mm long; glumes commonly surpassing florets; lemmas of staminate florets hairy, awnless; fertile lemma pubescent toward apex.

Most of Alaska. Moist meadows, marshes, tide flats, stream banks and less commonly on dry sites in arctic and alpine tundra. Common. Flowers May-July.

Hordeum brachyantherum   Meadow Barley

Tufted perennial grass to 85 cm tall. Ligule less than 0.5 mm long, leaf blade flat, mostly 2-6(7) mm broad, rough to touch, lacking auricles. Flower spike erect or nearly so, easily shattering, central spike sessile, pedicel of lateral spikelets about 1 mm long; glumes all slender, awn-like, mostly 7-12(15) mm long; central floret usually with an awn surpassing those of glumes, lateral florets usually reduced, often awn-like, sometimes staminate.

Coasts and islands of southern Alaska, less common in central interior and coastal western Alaska. Beaches, meadows and open areas. Common. Flowers June-August.

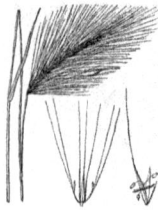

Hordeum jubatum (Foxtail Barley) Tufted, perennial, 25-65 cm tall; leaf blades flat, mostly 1-4 mm broad, rough to touch, lacking auricles or auricles present on some leaves; ligules less than 0.6 mm long; spike more or less nodding, 5-15 cm long (including awns), commonly almost as long as thick, easily shattering; central spikelet sessile; glumes slender, awn-like, mostly 30-100 mm long; central floret with awn often surpassing glumes; lateral spikelets usually reduced, often awn-like. Bogs, disturbed sites, beaches, roadsides, tide flats and river banks in most of Alaska south of 68th parallel. Flowers July-August.

Melica subulata　　　　Alaska Oniongrass

Tufted grass 30-90 cm tall, bulbous base atop short to elongate rhizomes; leaf sheath closed nearly or quite to top (sometimes split). Ligules 1.5-5 mm long, hairless; leaf blades flat, mostly 1-5(7) mm broad, usually rough to touch, at least on upper surface. Panicle 10-20 cm long, branches ascending to erect or sometimes spreading, spikelets 12-20 mm long, 2-5-flowered; glumes sharp-pointed, first about 5-6.6 mm long, second mostly 7.5-9 mm long; lemmas slender and sharp-pointed, unawned, more or less hairy on margins and nerves near base.

Islands of southern and southeastern Alaska. Meadows and woods. Uncommon. Flowers May-July.

Phalaris arundinacea　　　　Reed Canary Grass

Perennial strongly rhizomatous grass 80-160 cm tall. Ligules 4-9 mm long with minute hairs; leaf blades flat, 6-18 mm broad, with rough backward facing hairs. Panicles (8)11-20(30) cm long, compact or with branches more or less spreading; glumes almost equal, 4.0-5.6 mm long, more or less acute, usually wingless; fertile lemma 2.7-4.5 long, shining at maturity; sterile lemmas 2, 1.2-2 mm long, awl-shaped, hairy.

Southeastern, south-central and central Alaska. Stream banks, margins of streams, wet meadows and roadsides. Common. Flowers June-July. Native and Eurasian ecotypes are thought to exist in Alaska with a vast majority derived from non native ecotypes. Invasive populations are believed to be the result of crosses between cultivated and North American strains. It has been used as a forage grass and to stabilize road cuts. Excellent grass for basket weaving.

Phippsia algida   Snow Grass

Perennial, densely tufted grass 5-20 cm tall, stems erect or marginal ones spreading and bent alternately in different directions. Ligules 1-1.5 mm long, hairless; leaf blades flat, 1-2.5 mm wide or folded and less than 1 mm wide, hairless. Panicles congested, 0.8-4 cm long; spikelets 1.2-1.7 mm long; glumes unequal, much shorter than lemma and palea, membranous, sometimes lacking; lemmas hairless or sometimes sparingly hairy on nerves; anthers 0.3-0.6 mm long.

Southeastern, south-central, coasts and islands of western, west-central and northern Alaska. Uncommon. Flowers June-July.

Phleum alpinum   Alpine Timothy

Tufted grass 12-65 cm tall, stems sometimes rooting at lower nodes. Ligules 1-4 mm long; leaf blades flat or more or less folded to margins rolled under, 2-9 mm wide, more or less roughened at least on margins. Panicle 1-4.5(7) cm long, often more than 1 cm broad (when pressed); glumes fringed with hairs on keel and usually minutely pubescent on sides, awned, awn to 3 mm long; lemma minutely pubescent; anthers 1-1.6 mm long.

Southern half of Alaska. Thickets, stream banks, meadows and open woods. Common. Flowers June-August. Also known as *P. commutatum*.

KEY TO SPECIES OF *POA:*

1a. Stems stout, mostly 3-6 mm thick near base; spikelets mostly 6.5-11.5 mm long, ascending-erect, in stiff compact panicles; glumes almost as long as spikelet or surpassing florets; plants of ocean beaches and tidal flats.   *P. eminens*
1b. Stems various, seldom as much as 3 mm thick; spikelets variously arranged, sometimes 6.5-10 mm long, but then seldom in stiff compact panicles; glumes usually shorter than spikelet; plants of various distribution, seldom in tidal flats. **(2)**

2a. Plants annual, remains of old stems never present; anthers less than 1 mm long; second glume broadened near mid length; lemmas not webbed at base. *P. annua*
2b. Plants perennial, although sometimes flowering first season, usually with remains of stems from previous season; anthers usually more than 1 mm long, but if less, then second glume not broadened near mid length, lemmas webbed at base. **(3)**

3a. Anthers less than 1 mm long.   *P. leptocoma*
3b. Anthers 1-2.5 mm long or more. **(4)**

4a. Spikelets only slightly compressed to almost circular in transverse section, in flower more than twice (sometimes several times) as long as broad; lemmas rounded on back or only slightly keeled; plants of southeastern, south-central, and continental Alaska.   *P. scabrella*
4b. Spikelets compressed, in flower often less than twice as long as broad; lemmas more or less strongly keeled (except in *P. stenantha*, which has more or less soft-haired keel and marginal nerves and anthers less than 2 mm long); plants of various distributions. **(5)**

5a. Plants tufted; rhizomes lacking (stolons sometimes present). **(6)**
5b. Plants with rhizomes present, these frequently well developed. **(9)**

6a. Spikelets at flowering time about two-thirds as broad as long, almost heart-shaped; leaf blades flat, usually more than 2 mm wide; panicle pyramidal, about as long as broad; plants widespread.   *P. alpina*
6b. Spikelets at flowering usually less than two-thirds as broad as long, usually not almost heart-shaped; leaf blades often folded or margins rolled in from edges and less than 2 mm broad; panicle various. **(7)**

7a. Panicle open, more or less pyramidal; spikelets usually dark purplish; pubescence of marginal nerves usually softly long-villous; internerves usually pubescent towards base.   *P. arctica*
7b. Panicle more or less contracted, branches ascending to appressed or uncommonly more or less open; spikelets green or purplish, seldom dark purple; pubescence of marginal nerves with hairs not especially soft; internerves pubescent or hairless. **(8)**

8a. Lemmas hairy between keel and marginal nerves; plants mostly of coastal southern Alaska (less commonly inland).   *P. stenantha*
8b. Lemmas hairy only on keel and marginal nerves, internodes hairless or rarely minutely pubescent; plants widespread in Alaska.   *P. glauca*

9a. Spikelets mostly (5.7)6.0-10.4 mm long; plants mostly of coasts and islands of southern and less commonly interior Alaska.   *P. macrocalyx*
9b. Spikelets commonly less than 5.5 mm long; various distribution. **(10)**

10a. Stems and/or leaf sheaths with backward facing rough hairs; plants reported from southern Panhandle.   *P. laxiflora*
10b. Stems and leaf sheaths hairless, or rough to touch, not with backward facing hairs; plants widespread. **(11)**

11a. Lemma hairless or rough to touch between keel nerve and marginal nerve, not truly hairy on internerve area.   *P. pratensis*
11b. Lemma with soft or woolly interwoven hairs between keel nerve and marginal nerve, at least near base.   *P. arctica*

Poa arctica Arctic Bluegrass

Tufted or sod-forming rhizomatous, sometimes stoloniferous grass 10-50 cm tall, leaf sheaths hairless, stems smooth. Ligules 1.5-3.5 mm long, appearing gnawed or torn at edge, hairless; leaf blades folded, or flat, 1-2(3) mm broad, 1-2 stem leaves, upper often well below panicle. Panicle open, pyramid-shaped, 4-10(12) cm long (excluding drooping lower branches), branches spreading, or lowermost sometimes drooping; spikelets compressed, 4.5-6(7.5) mm long, 2-6-flowered, dark purple or greenish-purple, glumes unequal, shorter than adjacent lemmas, lemmas with long soft hairs on keel, marginal nerves, densely hairy to almost hairless on internerve, webbed or not at base; anthers 1.3-2.2 mm long.

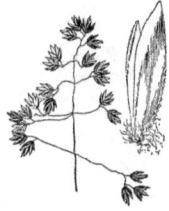

Most of Alaska. Tundra, woods, rock outcrops, stream banks, lake shores and meadows. Common. Flowers June-August.

Poa eminens Large-flower Speargrass

Perennial strongly rhizomatous grass 30-130 cm tall; leaf sheaths smooth, opaque; stems smooth, circular in transverse section, commonly 3-6 mm thick near base. Ligules 1.2-3 mm long, ciliate, hairless, leaf blades flat, 4-11 mm broad; stem leaves 2-4, uppermost placed near middle or upper part of stem. Panicles compact, (7)9-24 cm long, stiff branches appressed-erect or appressed-ascending; spikelets compressed, (6.5)7.0-11.5 mm long, 2-6-flowered, pale green maturing purplish or brownish; glumes almost equal to somewhat unequal, about equaling or surpassing adjacent florets and often surpassing all florets; lemmas with soft hairs on keel and marginal nerves, and more or less hairy on internerves, not webbed at base; anthers (1.6)2.3-4 mm long.

Coast and islands of southern and western Alaska. Beaches. Common. Flowers July-August.

Poa alpina (Alpine Bluegrass) Perennial, densely tufted, lacking rhizomes, 10-50 cm tall; leaf sheath hairless; stems smooth; ligules 2-4.4 mm long, hairless; leaf blades flat, less commonly folded, 2-4 mm broad, chiefly basal; stem leaves 1-2(3), uppermost near middle of stem or below in mature plants; panicle compact to more or less open, pyramid-shaped, 1.5-6(8) cm long, branches finally spreading; spikelets compressed, 4.5-7.7 mm long, mostly 3-6 flowers; glumes unequal, shorter than adjacent lemmas; lemmas hairy on keel and marginal nerves, more or less hairy on internerve, not webbed at base. Thickets, bogs, stream banks, lake shores, open woods and alpine and arctic tundra in most of Alaska. Flowers in July.

Poa annua (Annual Bluegrass) Annual, tufted, sometimes stoloniferous, 4-30 cm tall; leaf sheath hairless, thin transparent margin; stems smooth; ligules 0.7-3.2 mm long, appearing gnawed to entire, hairless; leaf blades flat or folded, 1-3.5 mm broad; stem leaves usually 1-2, uppermost commonly below panicle; panicles open, pyramid-shaped, 1.6-5(8) cm long, branches spreading; spikelets compressed, 3.8-6.5 mm long, 3-6-flowered, purple or purplish to green; glumes unequal; lemmas more or less hairy on keel and marginal nerves, not webbed at base. Disturbed sites in much of Alaska.

Poa glauca (Glaucous Bluegrass) Perennial, strongly tufted, not rhizomatous or stoloniferous, 10-60 cm tall; leaf sheath smooth to rough, opaque, those above basal tuft often purplish; stems smooth or more or less rough, circular in transverse section; ligules 1.2-3.8 mm long, hairless, stem leaves one or more, usually 2, upper one near middle of stem; panicles compact; spikelets more or less compressed, 2-6-flowered, green to purplish or dark purple; glumes almost equal to somewhat unequal, shorter than adjacent lemma, lemmas hairy on keel and marginal nerves, usually hairless on internodes, not webbed at base, or sometimes with a few folded hairs. Dry gravelly slopes, open woods, stream banks, rock outcrops in most of Alaska.

Poa leptocoma (Bog Bluegrass) Perennial, tufted, not rhizomatous, mostly 5-40 cm tall; leaf sheaths smooth or minutely rough-haired, opaque, sometimes lower ones purplish; ligules 1.3-3 mm long, entire or lance-shaped, hairless; leaf blades flat or folded, 1-4 mm broad, stem leaves 1-2; panicle compact, opening at maturity; spikelets compressed, 4.3-7.7 mm long, 2-5-flowered, purple or green turning purple; glumes unequal, shorter than adjacent lemmas, lemmas hairy on keel and marginal nerves, hairless on internerves, webbed at base. Alpine tundra in much of Alaska except Aleutians. Includes *P. paucispicula*.

Poa pratensis (Kentucky Bluegrass) Perennial, rhizomatous, sod-forming (seldom tufted), 25-100 cm tall; leaf sheath hairless or rarely with backward or forward facing rough hairs, opaque, often reddish above base; ligules 1-3 mm long, appearing gnawed, fringed with hairs, minutely pubescent; leaf blades flat or folded, 2-5 mm broad; stem leaves mostly 2-3(4); panicle open or compact; spikelets compressed, 4.0-6.5 mm long, 2-6-flowered, green to purplish; glumes more or less unequal in length and shape, shorter than adjacent lemma; lemmas more or less silky-hairy on keel and marginal nerves, strongly webbed at base. River banks, roadsides, lawns, bogs, spits, talus slopes, meadows and disturbed sites in most of Alaska. Flowers July-August. Indigenous and adventive in our region.

250

Poa stenantha (Trinius Bluegrass)
Perennial, tufted, 20-70 cm tall; leaf sheaths
smooth, opaque; ligules 1-3.5 mm long,
acute; leaf blade margins rolled under to flat,
0.5-2 mm broad; stem leaves commonly 1-3,
uppermost near middle; panicles narrow to somewhat open (5)7-15 cm
long; spikelets more or less compressed, keeled, (5.4)6-10 mm long,
3-5-flowered, stripped purplish; glumes unequal, much shorter than
adjacent lemmas; lemmas keeled, more or less soft hairs along keel
and marginal nerves, minutely pubescent to hairless on internodes, not
webbed at base. Thickets, marshes, roadsides, sea cliffs and rocky
outcrops on coasts and islands of southern Alaska and less commonly
into Interior.

Puccinellia nuttalliana   Shining Alkaligrass

Perennial tufted grass sometimes with runners that
take root or almost rhizomatous base, 40-70 cm tall,
ligules minutely hairy. Ligules 1.5-3 mm long, minutely
pubescent; leaf blades folded under or sometimes flat,
mostly 1-4 mm broad. Panicles 10-26 cm long, branches
appressed-ascending to spreading-ascending, rough to
touch; spikelets 7-13.5 mm long, 3-7(10)-flowered, green
to purplish, pedicels lacking enlarged tumid cells, glumes
minutely fringed with small hairs, first 1.3-2.4 mm long, second 2-3.8
mm long; lemmas 3.0-4.9 mm long, obtuse to somewhat acute, more
or less hairy towards base, palea usually shorter than lemma; anthers
(0.8)1-1.5 mm long.

Southeastern and south-central Alaska. Beaches and brackish
marshes. Common. Flowers June-July. Also known as *P. grandis*
and *P. lucida*.

Puccinellia nutkaensis   Pacific Alkaligrass

Perennial tufted grass to 70 cm tall. Ligules 1-3.5 mm
long, hairless, leaf blades flat to folded or margins rolled under,
1-3 mm broad. Panicles 4-19 cm long, branches appressed-
ascending to spreading, smooth or rough to touch, spikelets
5-8.5 mm long, (2)4-6(7)-flowered, green or purplish, pedicels
usually with enlarged tumid cells; glumes minutely fringed with
hairs, first 1.3-2.1 mm long, second 2-3.2 mm long; lemmas
3-3.8(4.9) mm long, somewhat acute, irregularly toothed to
lobed, more or less hairy, sometimes hairless at base; palea
almost equal to or sometimes surpassing lemma; anthers 0.7-
1.5 mm long.

Coasts and islands of southern Alaska. Beaches. Common
alkaligrass along coasts. Flowers June-August. Includes *P. hultenii*.

251

Torreyochloa pauciflora (Weak Mannagrass) Rhizomatous grass often rooting at lower nodes, 25-135 cm tall; leaf sheaths smooth or more or less rough, open; ligules 3-9 mm long; leaf blades flat, 3-15 mm broad, roughened on one or both sides; panicle mostly 10-23 cm long, rather loose; spikelets oblong to ovate in outline, compressed, 4.7-8 mm long, 3-7-flowered, first glume 0.8-1.8 mm long; lemmas 5-nerved or plainly 7-nerved, nerves often conspicuously rough to touch. Coasts and islands of southeastern and south-central Alaska in woods, thickets, marshes and meadows. Flowers June-August. Also known as ***Glyceria pauciflora*** and ***Puccinellia pauciflora***.

Puccinellia pumila (Dwarf Alkaligrass) Perennial, tufted, seldom stoloniferous, 8-20 cm tall; ligules 0.8-2 mm long, hairless; leaf blade margins rolled under, less than 1 mm broad; panicles 3-8(10) cm long, branches smooth; spikelets 4.5-5.5 mm long, 3-6-flowered, green or purplish, pedicels with enlarged tumid cells; glumes entire or more or less wavy; lemmas 3-3.5 mm long, obtuse to somewhat acute, not or only sparingly fringed with hairs, strongly nerved; palea usually shorter than lemma. Ocean beaches in southeastern and south-central Alaska.

Trisetum spicatum   Downy Oatgrass

Tufted grass to 70 cm tall, densely hairy with long soft hairs pointing down except at top. Ligules 0.8-1.5 mm long, fringed with hairs, leaf blades flat to folded, 1-5 mm broad, rough to touch. Panicles usually erect, compact, spike-like, mostly 1-13 cm long, branches very short, bearing spikelets from base, spikelets 5-7 mm long, 2-3-flowered, glumes unequal, first 3.8-5 mm long, second 4.2-6 mm long, not appearing gnawed near tip; lemmas 4.2-5.7 mm long, usually violet with brown margins; callus hairs about 0.5 mm long; awns bent and twisted, mostly 2-6 mm long.

Most of Alaska. Meadows, forests, alpine, rock outcrops, river banks and roadsides. Common. Flowers June-September.

Trisetum cernuum (Nodding Oatgrass) Tufted, often decumbent and more or less stoloniferous at base, 35-100 cm tall; ligules 1.5-3.7 mm long; leaf blades flat (3)4-8(10) mm broad, rough to touch; panicles nodding, open, 10-25 cm long, branches bearing spikelets toward tip; spikelets 7-9 mm long (excluding awns), 2-3-flowered; glumes unequal; lemmas 5-6.1 mm long. Woods, thickets, and marshes in coastal southeastern and south-central Alaska.

252

POTAMOGETONACEAE
PONDWEED FAMILY

Potamogeton natans Floating Pondweed

Floating and submerged perennial, leathery, linear leaves with many nerves, not or only slightly differentiated blade and petiole, stipules brownish or greenish, thick membranous free margins; fruit achene, green, shiny and wrinkled.

Lakes, ponds and streams in southeastern and south-central Alaska. Flowers May-August.

Potamogeton alpinus (Northern Pondweed)

Whole plant commonly tinged reddish, leaves mostly submerged, stalkless, stipules membranous with freed edges, not attached to leaf or shredded; fruit with curved beak. Common in lakes and sluggish streams in most of Alaska except northern coastal plain and Aleutians. Flowers July-August.

Potamogeton epihydrus (Nuttall Pondweed)

Stems somewhat flattened, leaves with a broad median band of air chambers, floating leaves conspicuously different from submerged ones, with dense flower spikes. Ponds and lakes in southeastern, central southern Alaska and western Aleutians. Flowers June-July.

Potamogeton gramineus (Grass-leaved Pondweed)

Narrow lance-shaped leaves, floating leaves usually present and differ from submerged leaves, stipules brownish, thick with membranous free margins. Ponds, lakes, and streams in most of Alaska south of Brooks Range. Flowers July-August.

Potamogeton vaginatus

(Sheathed Pondweed) Stems greenish or straw-colored, leaves all submerged, linear to threadlike, stalkless atop connected and sheathing brown stipules. Ponds, lakes and streams in most of Alaska. Flowers July-August. Also known as *Stuckenia vaginata*.

Zannichellia palustris

(Horned Pondweed) Submerged aquatic, branched stems from creeping rhizomes, leaves linear-thread-like, 1-nerved, acute, with sheathed, membrane-like stipules; staminate and pistillate flowers in same axils; fruit 2 mm long, 2-6 on short stalk with slender beak. Found in brackish water, common, often overlooked.

RUPPIACEAE
DITCH-GRASS FAMILY

Ruppia maritima   Ditch-grass

Aquatic perennial with thread-like stems to 80 cm long. Leaves submerged, linear to threadlike with membranous sheaths. Flowers in axillary spikes, long, spiraling peduncles, petals lacking. Fruit a pear-like drupelet about 2 mm long, tipped by straight style.

Southeastern Alaska west to eastern Aleutians. Salt or brackish water and tide flats. Common. Flowers July-August. Ducks feed on leaves, fruit and roots.

SCHEUCHZERIACEAE
SCHEUCHZERIA FAMILY

Scheuchzeria palustris   Scheuchzeria

Yellowish green aquatic rush-like perennial, mostly subterranean or submerged, yellowish gray creeping rootstock ending in a single, unbranched stem, 10-40 cm tall. Basal leaves erect, stem leaves reduced upwards with elongate ligule, sheathing at base, blade channeled or tubular. Flowers 3-12 in axils of well-developed bracts. Petals greenish-white. Fruit compressed follicles, seeds large, 4-5 mm long, light greenish-brown to dark-brown.

Southeastern, south-central and Interior Alaska. Sphagnum bogs, ponds and lake margins, usually fresh water. Uncommon. Flowers June-July. Also placed in *Juncaginaceae* and closely related to *Triglochin maritimum* (Arrowgrass).

SPARGANIACEAE
BUR-REED FAMILY

Sparganium emersum
Narrow-leaved Bur-reed

Aquatic perennial from rhizome with stems and leaves usually floating, 20-100 cm long. Leaves alternate, grass-like, alternate, upper leaves with dilated bases, distinctly nerved, elongate. Flowers borne in axils of leaves or with some along rachis. Achenes narrowed both ways from a swollen middle.

Much of Alaska south of Brooks Range. Deep or shallow water from lowland to alpine. Common. Flowers July-August. Also known as *S. angustifolium*.

Sparganium hyperboreum Northern Bur-reed

Aquatic herb with stems to 30 cm tall, floating or erect. Leaves erect or floating, thick and grass-like. Flowers in dense rounded clusters of 2-4 pistillate crowded heads, 1 stalkless staminate head. Fruit small, obovate, beakless with short stalkless stigma.

Most of Alaska. Shallow, fresh water to lower alpine. Common. Flowers July.

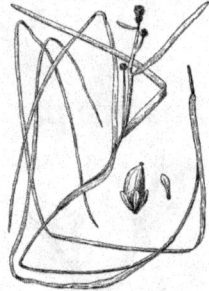

Sparganium minimum (Small Bur-reed) Aquatic, similar in size to *S. hyperboreum*, but staminate head peduncled and single, pistillate heads 2-3, widely separated, axillary. Ponds and lakes in southern southeastern, central western to eastern and south-central Alaska. Uncommon and may be confused with either *S. angustifolium* or *S. hyperboreum*. Flowers July-August. Also known as *S. natans*.

ZOSTERACEAE
EELGRASS FAMILY

Phyllospadix scouleri   Scouler's Surf-grass

Aquatic perennial with stout, flattened, usually unbranched stems from thick, fuzzy rhizomes. Leaves alternate, 2-ranked, linear, with 3 nerves. Inflorescence on short stalk, spathe dilated with a marginal appendage, lacking sepals and petals. Fruit heart-shaped with winged margins.

Islands of southeastern Alaska. Surf-beaten rocks in salt water, intertidal and subtidal zones. Occasional to common. Flowers June-July. Entire plant including long, salty leaves can be formed into cakes and dried for winter food. Leaves used to decorate baskets. Sometimes included in **Potamogetonaceae**. **Phyllospadix serrulatus** found in southeastern Alaska.

Zostera marina   Eel-grass

Aquatic perennial from slender rhizomes, stems branching and flattened. Leaves linear. Spadix enclosed by overlapping, entire margin of spathe, spathe leaf-like apically. Seeds ovoid and ribbed.

Southeastern Alaska west to Aleutians. Subtidal in protected water, less common in intertidal zone. Common. Flowers July-August. Rhizomes and leaf bases sweet eaten steamed or raw. Spawn of herring collected from leaves of **Zostera** and **Phyllospadix**. Damp leaves used to generate steam to bend cedar bentwood boxes.

GLOSSARY

Achene: A small, dry, one-seeded closed fruit in which ovary wall is free from seed.

Acuminate: Tapering to the apex, the sides more or less concave.

Acute: Distinctly and sharply pointed, but not drawn out.

Adnate: With unlike parts congenitally grown together.

Adventive: Imperfectly naturalized; not native.

Androgynous: Hermaphroditic; having both male and female flowers in the same inflorescence, with the male above the female.

Anther: The portion of the stamen that contains the pollen.

Annual: A plant that completes its life cycle in one year.

Annulus: A special ring of cells in fern sporangia which at maturity bursts the sporangia.

Apex: The tip, point, or angular summit of anything.

Apical: Pertaining to the apex or tip.

Appendage: An attached secondary part to a main structure.

Appressed: Lying flat against an organ.

Aquatic: Living in water.

Ascend: To go or move upward; to rise.

Attenuate: Long-tapering; the sides straight.

Auricle: An ear-shaped lobe or appendage.

Auriculate: With earlike appendages.

Awn: A bristle-like appendage, especially on the glumes of grasses.

Axil: The angle between a branch or leaf and the axis (main branch) from which it arises.

Axillary: Situated in the axil.

Basal: At or pertaining to the base.

Beak: A long, prominent projection; especially a prolongation of a fruit or carpel.

Bidentate: Having two teeth.

Biennial: Of two seasons' duration from seed to maturity and death.

Bifid: Two-cleft or 2-lobed.

Bilobed: Two-lobed.

Bipinnate: Twice pinnate, the pinnae again pinnate.

Blade: The expanded part of a leaf or petal.

Bloom: The white, waxy, powdery covering on many fruits, leaves, and stems.

Boreal: Northern

Bract: A small leaf from the axil of which a flower or a floral axis arises; also a small leaf just below a flower or flower cluster.

Bractlets: Bract borne on a secondary axis, as on a peduncle or petiole.

Bracteole: A bractlet or small bract.

Brackish: Salty.

Bulbets: A small bulb produced in the leaf axils, inflorescence, or other unusual places.

Bulbous: Having the character of a bulb.

Calcareous: Of or pertaining to calcium carbonate (limestone), as a calcareous soil.

Calyx: The outermost, usually green part of perianth of a flower; the sepals as a whole.

Callous: Having the texture of a callus.

Callus: A hard prominence or proturberence, specifically the hardened base of a spikelet of a grass.

Cambium: A layer, usually one cell thick, of persistent meristematic tissue, or a persistent meristematic layer that gives rise to secondary wood and secondary phloem.

Caudex: The woody base of a perennial plant.

Capillary: Hairlike; very slender.

Capitate: In heads; formed like a head; in a very dense or compact cluster.

Capsule: A dry, dehiscent (opening by definite pores or slits) fruit made up of more than one pistil.

Carpel: A simple pistil; one unit of a compound pistil.

Catkin: A scaly spike bearing apetalous, unisexual flowers; ament.

Chaff: Small membranous scales; degenerate bracts in many Asteraceae.

Channeled: Marked with one or more deep longitudinal grooves.

Ciliate: Said of a margin fringed with hairs.

Circumsessile: Opening or dehiscing along a horizontal line around the fruit or anther, the valve usually coming off like a lid.

Connate: United, used especially with reference to like structures.

Connective: That portion of the stamen that connects the two halves of an anther.

Corolla: The petals of a flower, collectively.

Corm: A solid erect bulblike stem with scalelike leaves, usually subterranean.

Corymb: Short and broad, more or less flat topped indeterminate flower cluster, the outer flowers opening first.

Corymbose: Arranged in corymbs.

Cotyledon: Seed leaf; the primary leaf or leaves in the embryo.

Culm: The jointed stem of grasses and sedges.

Cyme: A broad, flat-topped or convex flower cluster, with central flowers opening first.

Crenate: Said of a margin with rounded or blunt teeth.

Crustose: A type of lichen with a flat thallus that grows on rocks and tree trunks.

Decurrent: Extending down and adnate to the stem.

Decumbent: Reclining or lying on the ground, but with the end ascending.

Deflexed: Bent or turned abruptly downward.

Dehiscent: That which splits open, as the opening of anther or fruit along regular lines of suture.

Dichotomous: Branching by constantly forking in pairs.

Dimorphic: Occurring in two forms.

Dioecious: Unisexual, the male and female elements in different plants.

Dissected: Deeply divided or cut into many segments.

Disjunct: Said of a noncontinuous distribution of plant taxa.

Distinct: Separate; not united with parts in the same series.

Dorsal: Relating to the back or attached there; the surface turned away from the axis, which in a leaf is the lower surface.

Drupe: A fleshy, one-seeded indehiscent fruit, with seed enclosed in a stony endocarp called a pit.

Drupelet: One drupe in a fruit made up of aggregate drupes, as in a raspberry.

Elliptic: A flat part of a body that is oval and narrows to rounded ends, widest at or about the middle.

Elongate: Stretched; lengthened.

Emetic: An agent that causes vomiting.

Entire: Refers to margins of petals, sepals or leaves that are not toothed.

Epiphyte: A plant without connection to the soil, growing on another plant, but not deriving its food or water from it.

Exserted: Sticking out; projecting beyond, as stamens from perianth; not included.

Fascicle: A loose cluster or bundle of flowers, leaves, stems, or roots.

Feather Moss: Carpet-forming mosses on forest floor such as Rhytidiadelphus spp. Hylocomium splendens (Stairstep Moss), and Ptilium spp.

Fertile: Said of pollen-bearing stamens and seed-bearing fruits; capable of producing fruit.

Fibrous: Having numerous woody fibers; having the appearance of fibers, as the roots of monocots.

Fiddlehead: Immature fern in bud stage; forms a wound up ball.

Filament: The part of a stamen that supports the anther; threadlike structures.

Floret: The lemma and palea with included flower; also a small flower; one of a cluster.

Fluted: Having long parallel grooves in the trunk of a tree especially of Hemlock.

-foliate: A suffix meaning having leaves.

Foliage: The leafy covering, especially of trees.

Follicle: A single-carpellate dry fruit dehiscing along one line or suture.

Frond: The leaf of a fern including the stalk.

Gemma: A bud that develops into a new plant.

Glandular: Having or bearing secreting organs or glands.

Globular: Spherical.

Glume: One of the two bracts, found at the base of a grass spikelet, which do not subtend the flowers.

Glutinous: Covered with a sticky exudation.

Gynaecandrous: With staminate and pistillate flowers in same spike, the pistillate at the apex.

Habitat: The kind of locality in which a plant grows.

Hastate: Shaped like an arrowhead with the basal lobes turned outward.

Herbaceous: Not woody; dying to the ground each year; said also of soft branches before they become woody.

Humus: The earth, the soil, specifically that soil rich in decaying organic matter.

Hybrid: A plant arising from cross-pollination between two species of the same genus.

Hypanthium: A cup- or saucer-shaped enlargement or development of the receptacle that bear the calyx, corolla, and stamens at its apex.

Incised: Cut sharply and irregularly, more or less deeply.

Indehiscent: Not opening by valves or along regular lines.

Inflorescence: General distribution and arrangement of flowers on a stem.

Indusium: The epidermal outgrowth covering the sori on ferns.

Internodes: The part of the stem between two successive nodes.

Involucre: A cluster of bracts subtending a flower or inflorescence.

Irregular: Lacking in regularity of form; asymmetric, as a flower which cannot be halved in any plane, or one that is capable of bisection in one plane only.

Keel: A sharp or conspicuous longitudinal ridge; also the 2 partly united lower petals of many Apiaceae (Pea Family) species.

Lanceolate: Lance-shaped, rather narrow, tapering to both ends with the broadest part below the middle.

Lateral: On or at the side.

Lenticel: A slightly raised area in the bark of a stem or root.

Lemma: In grasses, the flowering glume; the lower of the two bracts immediately enclosing the flower.

Ligule: Flattened part of the ray corolla in Asteraceae and to the appendage on the inner (upper) side of the leaf at the junction of blade and sheath in many Poaceae and some Cyperaceae species.

Linear: Narrow and flat with sides parallel.

Liverworts: Green nonflowering plants related to mosses.

Lobe: A projecting segment of an organ.

Locule: A seed cavity (chamber) in an ovary or fruit.

Loment: A flat-seeded legume which is constricted between the seeds, falling apart at the constrictions when mature into one-seeded joints.

Margin: Edge of a leaf.

Membranous: Thin, more or less flexible and translucent, like a thin membrane.

Midrib: The main rib or central vein of a leaf or leaflike structure.

Midvein: Same as midrib.

Monoecious: With unisexual flowers, both sexes on the same plant.

Moraine: The accumulation or boulders, stones, and other debris deposited by a glacier.

Muskeg: A bog characterized by an abundance of sphagnum moss, more accurately called peatlands.

Mycorrhiza: The symbiotic relationship of a fungus and a plant root.

Naked: Lacking some structure, appendage, or hairs which might ordinarily be expected to be present.

Nectary: A nectar-secreting gland, often appearing as a protuberance, scale, or pit.

Nerve: In botany, a simple or unbranched vein or slender rib.

Node: The place on the stem where leaves or branches normally originate.

Nunatak: A high mountain peak that has remained above glacial ice.

Nutlet: A small nut.

Oblanceolate: Inversely lancelike, attached at the tapered end.

Oblong: Two to four times longer than wide and with nearly parallel sides.

Obovoid: A 3-dimensional figure reversed eggshaped in outline.

Obovate: Inversely eggshaped; attached at the narrow end.

Obsolete: Not evident or apparent; rudimentary; no longer used.

Obtuse: Blunt or rounded at the end.

Orbicular: Flat with a circular outline.

Ovary: The swollen basal portion of a pistil; the part containing the ovules or seeds.

Ovate: Eggshaped in outline, with the base broader than the tip.

Ovoid: A solid that is oval (less correctly, ovate) in flat outline.

Palmate: With three or more lobes, nerves, leaflets, or branches arising from a common point.

Palea: One of the pair of bracts (lemma and palea) which subtends the individual flowers in grass spikelets.

Panicle: A branched or compound raceme with two or more flowers on each branch with the younger flowers nearest the tip.

Paniculate: Having a panicle type of inflorescence.

Pappus: The modified calyx crowning the ovary (and achene) in Asteraceae.

Parietal: Borne on or pertaining to the wall or inner surface of an ovary or fruit.

Pedicel: Any slender stalk, especially one that supports a fruiting or spore-bearing organ. The stalk to a single flower of an inflorescence.

Peduncle: A flower-stalk supporting a cluster of flowers, or a single flower when the pedicel is very long.

Pedunculate: Borne on a peduncle.

Perennial: Living for more than two years and usually flowering every year.

Perianth: Term used when the calyx and corolla cannot be readily distinguished.

Perigynium A special bract which encloses the achene of Carex.

Perfoliate: Where the leaf has the stem apparently passing through it, or where opposite leaves are joined around the stem at their bases.

Persistent: Remaining attached after like parts ordinarily fall off.

Petiolate: With a petiole.

Petiole: The slender stem that supports the blade of a foliage leaf; a leafstalk.

Pinna: One of the primary divisions of a pinnately compound leaf.

Pinnate: Referring to pinnately compound leaves; leaves divided into leaflets arranged on two parallel sides of a common stem.

Pinnatifid: More or less deeply cut in pinnate fashion.

Pinnule: A secondary pinna or leaflet in a pinnately compound leaf.

Pistil: The seed-bearing organ of a flower, consisting when complete, of an ovary, style, and stigma.

Pistillate Having pistils and no stamens; female.

Pod: A dry dehiscent fruit.

Pome: A fleshy indehiscent fruit with an inferior ovary and more than one locule, as an apple.

Precocious: Appearing or developing very early.

Prostrate: Lying flat on the ground.

Pubescent: Covered with short soft hairs of down; loosely, bearing hairs.

Raceme: A type of simple inflorescence in which the individual flowers are borne on pedicels along a more or less elongated axis with the youngest flowers nearest the tip.

Rachilla: A diminuative or secondary rachis or axis.

Rachis: The central elongated axis to an inflorescence or a compound leaf.

Ray: The outer modified floret of some Asteraceae, with an extended or straplike flower axis on which some or all of the flower parts are borne.

Reflexed: Bent abruptly downward or backward.

Rhizomatous: Having the characters of a rhizome or bearing rhizomes.

Rhizome: A prostrate, elongated underground stem.

Riparian: Referring to stream and streamside vegetation.

Rootstock: Same as rhizome.

Rosette: A cluster of leaves attached at the base of plant near the ground.

Saline: Of or pertaining to salt.

Samara: A dry non splitting winged fruit.

Saprophyte: A plant deriving all of its nourishment from the bodies of decaying organisms.

Scale: Any thin, dry, membranous body, usually a degenerative leaf, sometimes of epidermal origin.

Scape: The leafless flowering stem of a plant; it may bear scales or bracts but not foliage leaves.

Scapose: Bearing a scape or resembling one.

Scorpoid: Said of a coiled cluster in which the flowers are two-ranked and borne alternately at the right and the left.

Sepal: One of the parts of the outer whorl of a floral envelope or calyx, usually green in color.

Septate: Partitioned; divided by one or more partitions.

Septicidal: A capsule splitting down the septa and not through the locule.

Sessile: Without a stalk.

Seta: A bristle or bristle-shaped body.

Sheath: A tubular envelope, usually used for that part of the leaf of a sedge or grass that envelopes the stem.

Silicle: Short fruit of certain species in the mustard family (Brassicaceae) usually less than 5 times longer than broad.

Siliques: A long narrow, many seeded capsule of certain species of the mustard family (Brassicaceae), with two valves usually separating at maturity; usually more than 5 times longer than broad.

Sinus: That space or recess between two lobes or divisions of a leaf or other expanded organ.

Sori: A cluster or grouping of spore-containing bodies on a fern frond.

Spadix: A spike with a thick and fleshy axis, usually densely flowered with imperfect flowers.

Spathe: A large bract enclosing an inflorescence in the Arum lily family.

Spike: An inflorescence consisting of a central rachis bearing a number of sessile flowers.

Sporangium: A structure within which spores are produced.

Spore: A cell which becomes free and capable of direct development into a new individual; the first cell of a gametophyte.

Sporophyll: A spore-bearing leaf, often highly modified.

Spreading: Diverging nearly at right angles; nearly prostrate.

Stamen: The male organ of a flower, consisting of an filament and an anther, the latter bearing the pollen.

Staminate: Having stamens and no pistil; male.

Staminoidia: A sterile stamen; lacking an anther.

Sterile: Infertile and unproductive, as a flower without a pistil, a stamen without an anther or a leafy shoot without flowers.

Stigma: The apex of the pistil; the part that receives the pollen.

Stipe: The stalklike support of a pistil; also the name of the petiole of a fern frond.

Stipitate: Provided with a stipe or a slender stalklike base.

Stipulate: Provided with stipules.

Stipule: An appendage at the base of a petiole or leaf or on each side of its insertion.

Stoloniferous: With stolons or runners that take root.

Stolon: A trailing shoot above ground rooting at the nodes; a runner.

Stomates: A small opening on the surface of a leaf through which gaseous exchange takes place.

Strobilus: A structure characterized by partly overlapping bracts or scales, as a pine cone.

Subtend: To stand below and close to, as a bract below a flower or a leaf below a bud.

Subshrub: A very low shrub loosely treated as a perennial.

Subterranean: Below the surface of the ground.

Succulent: Fleshy and full of juice.

Style: A prolongation of the ovary commonly bearing the stigma.

Subshrub: A low, woody plant that is botanically a shrub, but for this guide classified as an herb due to its small size and presence in the herb layer.

Talus: A usually bouldery or gravelly slope.

Taproot: The primary descending root, forming a direct continuation of the stem.

Terminal: Proceeding from, or belonging to the end or apex.

Ternate: In threes.

Thallus: A plant body which is not clearly differentiated into roots, stems and leaves.

Tomentose: With densely woolly, soft, matted hairs.

Transverse: Across; at a right angle to the longitundinal axis.

Tridentate: Three-toothed.

Tuber: A short, thickened branch of a subterranean stem, beset with buds or "eyes".

Tuberoids: A fleshy, thickened root resembling a tuber.

Trifoliolate: Having three leaflets.

Truncate: Ending abruptly, the base or apex nearly or quite nearly straight across.

Tufted: Having a cluster of hairs or other slender outgrowths; stems in a very close cluster.

Turions: A sucker or root emerging from the ground.

Tussocks: A clump or tuft of grass or sedge.

Umbel: A racemous inflorescence with very short axis and more elongate pedicels which seem to arise from a common point.

Utricle: A small, thin-walled, 1-seeded, more or less inflated fruit.

Whorl: An arrangement of one or more leaves at one node.

Witches Broom: Enlarged branch of western hemlock and other species caused by the growth of dwarf mistletoe which parasitize trees in the pine family by sending its roots into the phloem of the host species to obtain nutrients.

REFERENCES

Abrams, L. 1960. Illustrated Flora of the Pacific States: Washington, Oregon, and California. 4 vols. standford Univ. Press, Stanford, CA.

Anderson, H.P. Flora of Alaska and Adjacent Parts of Canada. Iowa State University Press. 1959 543pp.

Brayshaw, T.C. 1976. Catkin bearing plants of British Columbia. Occas. Pap. of British Columbia Prov. Museum No. 18. Prov. British Columbia, Victoria, B.C.

Calder, J.A. and R.L. Taylor. 1968. Flora of the Queen Charlotte Islands Part 1. Canada Department of Agriculture, Research Branch. Monograph No. 4. Part 1. 659 pp.

Coupe, R, C.A. Ray et al. 1982. A Guide to Some Common Plants of the Skeena Area, Britsh Columbia. Ministry Of Forests. Province of British Columbia.

Douglas, G.W. 1982. The Sunflower Family (Asteraceae) of British Columbia. British Columbia Provincial Museum. Victoria, Canada.

Graham, F.K. and the Ouzinkie Botanical Society. Plant Lore of an Alaskan Island. Alaska Northwest Publishing Company. Anchorage. 1985.

Jacques, D. 1973. Reconnaisance botany of alpine ecosystems on Prince of Wales Island. MS Thesis, Oregon State University, Corvallis, OR. 133 pp.

Jones, A. Nauriat Niginaqtuat: Plants That We Eat. Written under a grant from the Indian Health Service. 1983

Hitchcock, A.S. 1971. Manual of the Grasses of the United States. Dover. New York. Reprint of original 1935 edition.

Hitchcock, C.L., A. Cronquist, M. Ownbey, and J.W. Thompson. 1955-1969. Vascular Plants of the Pacific Northwest. University of Washington Press, Seattle.

Hitchcock, C.L. and A. Cronquist. 1969. Flora of the Pacific Northwest. University of Washington Press, Seattle. Five Vols.

Hultén, E. 1968. Flora of Alaska and neighboring territories. Stanford University Press. Stanford, CA. 1068 pp.

Kartesz,J.T. and R. Kartesz. 1980. A synonymized checklist of the vascular flora of the United States, Canada and Greenland. Vol II The Biota of North America. The University of North Carolina Press, Chapel Hill. 498 pp.

Klinka,K., V.J. Krajina, A. Ceska and A.M. Scagel. 1989. Indicator plants of coastal British Columbia. University of British Columbia Press, Vancouver. 288 p.

MacKinnon, A.; J. Pojar; R. Coupe. 1992. Plants of Northern British Columbia. British Columbia Ministry of Forests and Lone Pine Publishing.

REFERENCES

Martin, J.R. 1989. Vegetation and environment in old growth forests of northern SE, Alaska: a plant association classification. M.S. thesis, Arizona State University, Tempe, AZ. 221 p.

Murray, D.F. and R. Lipkin. 1987. Candidate threatened and endangered plants of Alaska. University of Alaska Museum, Fairbanks, AK. 75 pp.

Muller, M.C. 1982. A preliminary Checklist of the Vascular Plants in Southeastern Alaska. Wildlife and Fisheries habitat management Notes No. 5. USDA Forest Service Alaska Region Admin. Doc. No. 112. Juneau, AK. 32 pp.

O'Clair, R; R.H. Armstrong; R. Carstensen. 1992. The Nature of Southeast Alaska. Alaska Northwest Books, 254 pp.

Perkins,G.K. 1982. Botanical studies on Prince of Wales Island, Alaska. Ph.D. diss. Mississippi State University, Mississippi State, Mississippi. 146 pp.

Scoggan, H.J. 1978-1979. The Flora of Canada. Publications in Botany No. 7 (1-4). National Museums of Canada, Ottawa.

Smith, J.P. 1977. Vascular Plant Families. Mad River Press. 320pp

Stone,C. 1983. Patterns in coastal marsh vegetation of the Juneau Area, Alaska. Ph.D Diss. Oregon State Unversity, Corvallis, OR. 259 pp.

British Columbia Provincial Museum. Victoria, Canada Publications:
Heather Family of British Columbia. Szczawinski, A.F. 1962.
The Orchids of British Columbia. Szczawinski, A.F. 1959.
Common Edible Plants of British Columbia. 1962
Ferns and Fern Allies. Taylor, T.M.C. 1956.
Figwort Family of British Columbia. Taylor, T.M.C 1974
Lily Family of British Columbia. Taylor, T.M.C 1974
Pea Family of British Columbia. Taylor, T.M.C 1974
Sedge Family of British Columbia. Taylor, T.M.C 1983

Taylor, R.L., B. MacBryde. 1977. Vascular plants of British Columbia: a descriptive resource inventory. Tech. Bull. 4, The Botanical Garden, University of British Columbia. University of British Columbia Press, Vancouver. 754 pp.

Viereck, L.A.; Little, E.L. 1972. Alaska Trees and Shrubs. U.S. Department of Agriculture. Forest Service. Agricultural Handbook No. 410.1972. 265pp.

Welsh, S.L. 1974. Anderson's Flora of Alaska and Adjacent Parts of Canada. Brigham Young Univ. Press, Provo, Utah. 724 pp.

INDEX

269

INDEX

272

FIELD NOTES

Native Plants of Southeast Alaska is available at many bookstores in Alaska. If you cannot purchase it from a bookstore, check your local library or order directly from the publisher.

Please send _____ copies of
Native Plants of Southeast Alaska
$19.99 each & $4.50 S&H to:
Name_____
Address: _____
City_____State_____
Zip_____
Contact info: judyinhaines@aptalaska.net
Judy Hall Jacobson
Windy Ridge Publishing
P.O. Box 1158, Haines, AK 99827

www.ingramcontent.com/pod-product-compliance
Lightning Source LLC
Chambersburg PA
CBHW072112270326
41931CB00010B/1534